Power in the Wild

*The Subtle and
Not-So-Subtle Ways
Animals Strive for
Control over Others*

POWER
in the Wild

LEE ALAN DUGATKIN

The University of Chicago Press
Chicago and London

The University of Chicago Press, Chicago 60637
The University of Chicago Press, Ltd., London
© 2022 by Lee Alan Dugatkin
All rights reserved. No part of this book may be used or reproduced in any
manner whatsoever without written permission, except in the case of brief
quotations in critical articles and reviews. For more information, contact
the University of Chicago Press, 1427 E. 60th St., Chicago, IL 60637.
Published 2022
Printed in the United States of America

31 30 29 28 27 26 25 24 23 22   1 2 3 4 5

ISBN-13: 978-0-226-81594-7 (cloth)
ISBN-13: 978-0-226-81595-4 (e-book)
DOI: https://doi.org/10.7208/chicago/9780226815954.001.0001

Map illustration by Katie Shepherd Christiansen.

Library of Congress Cataloging-in-Publication Data

Names: Dugatkin, Lee Alan, 1962– author.
Title: Power in the wild : the subtle and not-so-subtle ways animals strive
    for control over others / Lee Alan Dugatkin.
Description: Chicago : University of Chicago Press, 2022. | Includes
    bibliographical references and index.
Identifiers: LCCN 2021021572 | ISBN 9780226815947 (cloth) |
    ISBN 9780226815954 (ebook)
Subjects: LCSH: Social behavior in animals. | Social hierarchy in animals. |
    Decision making in animals. | Aggressive behavior in animals.
Classification: LCC QL775 .D845 2022 | DDC 591.56—dc23
LC record available at https://lccn.loc.gov/2021021572

♾ This paper meets the requirements of ANSI/NISO Z39.48-1992
(Permanence of Paper).

CONTENTS

A gallery of photos follows page 104.

ELEPHANT
SEALS

LOONS

CARIBOU

FALLOW
DEER

SKYLARKS

HOF

WASPS

SCRUB JAYS

SWORDTAILS

RHESUS
MACAQUES

HOWLER
MONKEYS

CICHLIDS

CAPUCHINS

ARGENTINE
ANTS

VENS

DOGS

GELADA
BABOONS

CHIMPS

RED-TAILED
MONKEYS

WHITE-FRONTED
BEE-EATERS

PIGTAIL
MACAQUES

ONOBOS

MONGOOSES

HYENAS

CICHLIDS

OLIVE BABOONS

LEMURS

ANGELFISH

ERKATS

DOLPHINS

CUTTLEFISH

SUPERB FAIRY
WREN

LITTLE BLUE
PENGUINS

Species	Location
Hyenas	Masai Mara Reserve, Kenya
Lemurs	Kirindy Mitea National Park, Madagascar
Chimps	Budongo Forest Reserve, Uganda
Meerkats	Kuruman River Reserve, South Africa
Olive baboons	Gombe National Park, Tanzania
Bonobos	Wamba Territory, Democratic Republic of the Congo
Gelada baboons	Amhara Region, Ethiopia
Cichlids	Lake Tanganyika, Burundi
White-fronted bee-eaters	Gilgil, Kenya
Mongooses	Queen Elizabeth National Park, Uganda
Red-tailed monkeys	Kibale National Park, Uganda
Loons	Rhinelander, Wisconsin
Elephant seals	Santa Cruz, California
Wasps	Monroe, Louisiana
Argentine ants	Río Paraná de las Palmas and Rio Uruguay, Argentina
Scrub jays	Archbold Biological Station, Venus, Florida
Caribou	Grands-Jardins National Park, Lac-Pikauba Unorganized Territory of Canada
Swordtails	Veracruz, Mexico
Cichlids	Lake Xiloá, Nicaragua
Capuchins	Barro Colorado Island, Panama
Rhesus macaques	Cayo Santiago Island, Puerto Rico
Howler monkeys	Los Tuxtlas, Mexico
Skylarks	Orsay, France
Ravens	Grünau im Almtal, Austria
Horses	Camargue, France
Fallow deer	Dublin, Ireland
Pigtail macaques	Perak, Malaysia
Dogs	Kolkata, India
Angelfish	Lizard Island, Australia
Cuttlefish	Whyalla Bay, Australia
Superb fairy wrens	Canberra, Australia
Little blue penguins	Banks Peninsula, New Zealand
Dolphins	Shark Bay, Australia

PREFACE

The Wolf Science Research Park is about an hour's drive due north of Vienna. In the winter of 2018, I was visiting there to give a talk on my book *How to Tame a Fox and Build a Dog*. That book is all about an ongoing 62-year-long experiment studying domestication in silver foxes in real time in Novosibirsk, Russia. From the start of that experiment, Lyudmila Trut, my coauthor on the book, and her colleagues have been especially interested in dog domestication. For a whole suite of reasons—scientific, logistical, and political—they have been using the silver fox (*Vulpes vulpes*) as a stand-in for the wolf to study, step by step, how domestication occurs. So the chance for me to spend some time up close and personal with wolves was, to say the least, exciting.

Shortly after I gave my lecture, Kurt Kotrschal, my host and the director of the park, took me on a tour of the facilities, which house a half dozen or so wolf packs, each in its own outdoor enclosure. Kurt and his colleagues raise every wolf from birth, and each wolf knows these humans well. His research group has studied many aspects of the social behavior of these wondrous creatures—everything from feeding and choice of mates to cooperation (with other wolves as well as with humans), play behavior, exploration, fear, dominance, and power. The wolves never fail to teach Kurt's team something new.

It's a large research park, and each pack of wolves has its own fenced-off territory. There are also indoor facilities where the wolves are weighed and measured and experiments are run. As Kurt and I passed through the gate into the territory of one pack, he told me the standard operating procedure: when a wolf approaches, I was to get down on my knees, not make any sudden moves, and, as Kurt made very clear, "engage in friendly eye contact, but don't stare — just like with humans." The "not make any sudden moves" directive was not especially reassuring, but I trusted that Kurt knew these animals as well as any human can.

A large older male wolf approached us, and Kurt fed him a few pieces of food that he kept in his pocket for just such occasions. That wolf then ambled over to me — on my knees as instructed — lifted his paw and, for a brief instant, placed it on my shoulder. It was simultaneously terrifying and wonderful, and I sat there in awe, thinking, "That is one powerful animal. He could clearly kill me, if he so desired."

That's one component of power: strength. But it's a rather mundane, not especially interesting, component. A few minutes later, I was treated to a more dynamic, intriguing way that power manifests itself. As Kurt and I walked past a different pack, most of the animals were going about their daily routine, which, for the most part, meant lazing about, doing nothing. But the two wolves nearest us were engaged in rather different business. One was sitting on top of the other, with its jaws clamped down on the snout of the unfortunate below it. I was clearly taken aback, and Kurt, sensing my discomfort, told me that the dominant male was not harming the subordinate in its lock. It was performing a display of power, an "I'm in charge here" display, he told me, meant for the subordinate and perhaps for Kurt and me as well.

As we moved along the gravel road that runs through the Wolf Science Research Park, Kurt and I came upon another set of large fenced-off enclosures. Each housed a pack of dogs, all raised by humans. The dogs, like the wolf packs, were free to do as they pleased with Kurt and his team doing what they could to record their daily lives and run experiments on them, as they were doing with the wolves. Indeed, one day they hope to get some of Lyudmila's domesticated foxes and raise them by hand like the wolves and dogs, so that they can compare power dynamics, and many other things, across all three canine species.

The quest for power played out differently in those dogs than in the wolves across the road. It seemed, and Kurt confirmed this, that their interactions were a bit more chaotic, that attaining power revolved around petty, aggressive, and, on rare occasions, dangerous scrums to rise up to the top of the heap.

For the last thirty years, I've been studying many aspects of the evolution of social behavior. When I started in graduate school, I had honed down my list of dissertation topics to either the evolution of cooperation or the evolution of dominance. In time, it became clear to me that these topics are not mutually exclusive, but I did see them that way back in 1988 and felt that I needed to make a choice. I settled on the evolution of cooperation, mostly because the field of animal behavior was abuzz with ideas about cooperation and altruism at that time. But it was a very close call, and my passion for understanding dominance led to a few side projects on that topic as I worked on my dissertation. I added studies on cultural transmission in animals to those I was doing on cooperation and dominance, and over time I realized that social behavior in all these areas involved subtle, nuanced assessments and decisions. So many of those social decisions, like those Kurt's wolves and dogs were making, revolved around the ability to direct, control, or influence the behavior of others and/or the ability to control access to resources: what I am defining throughout this book as power. That realization was a cathartic experience for me, and judging from the published literature in animal behavior, I'm not alone. Right now, all over the planet, the work on power in nonhumans has become so cutting edge, so exciting, and so replete with adventure, that it's time to tell its story.

Power—or more specifically the quest to attain and maintain power—lies at the heart of almost all animal societies. The subtle, and often not-so-subtle, ways that animals seek power over those around them are astonishing and informative, both in and of themselves and because they provide an evolutionary window through which we can better understand behavioral dynamics in group-living species.

In these pages you'll discover that animal behaviorists (also called ethologists), psychologists, anthropologists, and other scientists have come to realize that power pervades every aspect of the social lives of animals: what they eat, where they eat, where they live, whom they mate with, how many offspring they produce, whom they join forces with, whom they work to depose, and more. Sometimes power struggles are between males, sometimes between females, and sometimes across sexes. At times, power pits young against old; at other times, the struggle is mostly with peers. Sometimes kin are pitted against one another, and other times they join forces to usurp the power of others.

With so much at stake, the quest for power may involve overt aggression, but many times it entails the use of more nuanced strategic behaviors: complex assessments of potential opponents, spying, deception, manipulation, formation of alliances, and the building of social networks, to name just a few. What's more, researchers have developed theories to understand the evolution of those strategic paths to power, and have derived and tested predictions generated by those theories, both in the field and in the laboratory. Understandably, much of that work focuses on behavior per se, but we'll also be getting glimpses of the hormones, genes, and neural circuitry underlying power.

Chapter 1 will provide an overview of the incredible ways that power manifests itself in nonhumans, while chapter 2 will examine the costs and benefits that drive the evolutionary trajectories that power takes. After the basic framework in these chapters is in place, we will explore how animals assess one another in the struggle for power (chapter 3). In so doing, we will dive into the myriad ways that nonhumans employ information from their own experiences, as well as what they have gleaned by watching (and being watched by) others, in part to form alliances critical to the struggle for power (chapters 4–5) and to cement their hold on power should they attain it (chapter 6). From there, we will explore how and why, in some species, group dynamics play an important role in building power structures (chapter 7). Finally, in chapter 8, we will see that while power structures are often stable, sometimes they crumble, only to be rebuilt in new ways. The species included in each chapter are not the only ones that inform our understanding of power, but rather the ones particularly well suited to do so.

Power is multidimensional, which means that some of the case studies we examine could be discussed in more than one chapter. On occasion, as in the endlessly fascinating power dynamics in hyenas, ravens, and dolphins, I return to a system across chapters. More often, though, because there are so many incredible systems from which we can learn about nonhuman power, I restrict discussion of a given system to a single chapter, basing that (admittedly somewhat subjective) decision on where I think it works to best to shine a light on power.

All of which is to say that our journey to understand the quest for power in nonhumans will require casting aside any notion that it is simple and straightforward. It isn't.

No one knows that better than the scientists who are studying power in animals. I've talked extensively with the researchers behind every system we will examine. Though history has not always acknowledged their role, here I make a concerted effort to see that the work of female scientists studying power receives the attention it so richly deserves. Close to 40% of the studies we will cover were led (or co-led) by women. I also made a conscious effort to include the work of younger, as well as more seasoned, researchers.

All the scientists I contacted were not just open to my barrage of personal and scientific questions, but also remarkably generous with their time. How these fascinating people came to do their studies, the day-to-day work involved, the twists and turns, the serendipity and the bad luck as well, all provide a narrative backdrop for their work on power. We'll see what would lead a researcher to lie in mounds of guano for hours on end in the middle of the night to observe penguin power plays in New Zealand, how a vacation outing to the Ngorongoro Crater in Kenya over thirty years ago led to an incredible study of hyena power that is still ongoing, how a chance encounter at a garage solved a problem and helped forward our understanding of spies and power in swordtail fish, and how a childhood love of the street dogs of Kolkata eventually turned into one of the most detailed analysis to date of social behavior and animal power in an urban setting.

The quest for power is evident everywhere, and in every sort of animal imaginable. You will see power dynamics in animals including hyenas, meerkats, mongooses, caribou, chimpanzees, bonobos, macaques, baboons, dolphins, deer, horses, and field mice, as well as ravens, skylarks, white-fronted bee-eaters, common loons, Florida scrub jays, copperhead snakes, wasps, ants, and cuttlefish. In each case, we will explore why that species and why that locale, the dynamics of power, the hypotheses being tested, and how and why scientists tested those hypotheses. We will travel with these researchers to the bays and the botanic gardens of Australia; the forests of the Democratic Republic of the Congo, Tanzania, Uganda, and Panama; the streets of Kolkata and Southern California; the meadows of the South of France; the parks of Dublin; the lakes of Michigan, Wisconsin, and Nicaragua; the mountains of Austria; the tundra of Canada; the beaches of New Zealand; the reserves and cliff faces of Kenya and more—all to understand the hows and whys of power in nonhuman animal societies.

The word "nonhuman" is key; this is not a book about power in humans. Evolutionary anthropologists and others have written much about the evolution of power in our own species. Indeed, we don't need to reference human behavior to appreciate the meaning of power in animals: in that sense, this book is a stand-alone tribute to the complexity, the depth, and, dare I say, the beauty of power in animal societies.

1   Chart a Path to Power

...the hyena...sad yowler, camp-follower, stinking, foul...
ERNEST HEMINGWAY, *The Green Hills of Africa*

If Hemingway was implying that hyenas are boorish, brutish, stupid beasts, he got it wrong.[1] Masai herders put cowbells around the necks of their cattle, and spotted hyenas (*Crocuta crocuta*) are savvy enough to tell the difference between the sound of those bells and the sound of church bells. Spotted hyenas also recognize all members of their groups, or clans, by sight and sound, cooperate when defending group territories and when hunting, and raise their offspring communally. Their paths to power—paths that are primarily traversed by females—are equally complex. What makes the hyena system even more intriguing is that, in an exception to the rule for mammals, adult female hyenas always outrank males within clan hierarchies.[2]

No one knows all of this better than Kay Holekamp. Not that she ever imagined she would be studying such things, as her PhD work at the University of California, Berkeley, was on behavior and dispersal in Belding's ground squirrels living in the Sierra Nevada. But in 1976, she and her then-husband decided an adventure was in order, so they saved up the money and went to Kenya on holiday. While they were visiting Ngorongoro Crater, they spotted a pack of hyenas chasing and hunting down, in a coordinated fashion, a wildebeest.

"They managed to take this wildebeest down and tear it to pieces right next to our vehicle," Holekamp recalls. "I turned to Rick and said, 'I thought those things were supposed to be skulking carrion eaters, not coordinated, cooperative hunters.'" When they returned stateside, she read ecologist Hans Kruuk's book on hyenas and became even more enamored with these creatures. Along with Laura Smale, in 1988, Holekamp began a field study of spotted hyenas at Masai Mara Reserve in Kenya. Over more than three decades, the study has grown to involve more than a hundred students and collaborators from around the world.[3]

Masai Mara sits at an elevation of 1,500 meters. Its grasslands—with their resident gazelles, lions, leopards, topi, and migrating zebras and wildebeest—surround Kay Holekamp's Fisi (Swahili for hyena) basecamp, with its tents, tables, and liquid nitrogen tanks for freezing blood. From Holekamp's home away from home at Fisi, she and her team study social dynamics in hyena clans to unravel the mysteries of this unusual, hyper-social creature. Most clans have about 50 individuals, but one, the Talek West clan, is an exception, numbering 130 or so members. Communal dens—underground labyrinths, often in the remains of an aardvark's haunt—are used by females when raising pups. On occasion, there are skirmishes between adults in different dens, as when there was a bit of a brouhaha between members of Dave's Den, the Lucky Leopard Den, and the Mystery Den.

Holekamp and her team can recognize each animal by its unique fur pattern. They collect anything and everything that might shed light on hyenas. On occasion, hyenas are darted with the tranquilizer Telazol, and their weight and size are recorded. Blood samples, rich with data on hormones, are drawn, and anal swabs are gathered, to further delve into hyena hormones. Many adult hyenas, who generally tip the scales at 100–150 pounds and stand about 3 feet tall at the shoulders, have been fitted with radio collars by which their GPS coordinates can be monitored, so that not only their locations, but their proximity to others in their clan, can be continuously recorded.

Besides relying on the GPS data, Holekamp and her team are

always on the move in Land Rovers observing hyenas, constructing an account of their moment-by-moment interactions. Their database includes thousands of detailed observations on power-related behaviors, including "ears back," a submissive act displayed by subordinates when they are threatened by a dominant hyena; "open-mouth appease," a behavior that appears to preempt aggression, in which one hyena presents an open-mouth gesture to another; and "stand over," in which a dominant individual reinforces its rank by keeping its head high and its muzzle facing down, poised above the shoulders of a subordinate. Back at Michigan State University, Holekamp's academic home, she went so far as to construct a life-size hyena robot, with a built-in camera and tape recorder, capable of reproducing many of these behaviors—it can pull its ears back, open its mouth, move its head up and down, and more. So far, while it makes for quite the conversation starter, Holekamp has not figured out a way to bring the hyenabot over to Africa and employ it to experimentally manipulate power-based interactions in a clan. That said, what she and her team know is that when a hyena rises to the top of the clan power structure, the benefits can be substantial. Those benefits can be especially significant for females, as they outrank males. Holekamp and her team wanted to know why females are usually dominant over males. It turns out it's because of their massively strong, bone-crunching jaws.

You don't want a hyena clamping its jaws down on you: their skulls, jaws, and teeth together can crack open the bones of zebras and giraffes. Ninety-five percent of the carcasses eaten by hyenas are the result of fresh hyena kills, not scavenging, and when they feast on their kills, the better part of which may be gobbled up in a matter of minutes, there is intense competition over the meat.

It takes a long time to develop a skull and jaws strong enough to crack bones that can measure more than 3 inches in diameter. This fact makes hyena pups reliant on their mothers, who are the only parents providing care, for much longer than in closely related species. The upshot of this is that there is especially strong natural selection pressure on females favoring a jaw capable of cracking open the bones of a fresh carcass. These bone-crusher jaws allow adult females to feast on a carcass themselves and to give their pups, who join their mothers at carcasses when they are about three months old, preferential access to kills. They are also an effective weapon in the struggle for power, and that, in part, explains why females outrank males.[4]

High-ranking mothers tend to produce daughters who also rise high in the clan power structure. What's more, the most powerful females—those sitting atop the dominance hierarchy—reap the gold standard of evolutionary benefits: increased long-term reproductive success. In a seven-year study of one clan in the Talek area of Masai Mara, Holekamp and her team not only collected data on social rank, but amassed a database on dozens of females from fourteen families, replete with information on litter sizes, time between births, how old pups were at the age of weaning, and how likely pups were to reach the age of reproduction themselves.[5]

High-ranking females began reproducing at a younger age than their lower-ranking counterparts, resulting in a 10% longer reproductive life span. The powerful gave birth more often (had shorter interbirth intervals), and they produced pups that were more likely to survive to the age of reproduction themselves, creating dynasties of a sort within hyena clans. The path to hyena power, when successfully navigated, pays.

The hyena way is just one path to power. The path that northern elephant seals traverse is different, with full-fledged knock-down, drag-out fights between males, and with females pitting male against male when need be. The cuttlefish path to power is a bit more subtle, using color, pattern, and deception, though they are not at all averse to turning to aggression when necessary. Swordtails take the subtle approach to power a step further and spy on potential rivals.

After our initial sortie into the paths to power in elephant seals, cuttlefish, and swordtails in this chapter, we'll weave together work from across the globe, drawing on cutting-edge studies in animal behavior, evolution, economics, psychology, anthropology, genetics, endocrinology (the study of hormones), and neurobiology in our quest to understand power in animal societies. In so doing, we will delve deep into how, in their quest for power, animals strategically weigh the costs and benefits of their power-related actions, assess their rivals, eavesdrop on others, build alliances (and even super alliances) to climb the power ladder, cement their hold on power, manage power at the level of the group, and rise through the power structure or get deposed. In each case, we'll not only delve into the science, but hear the behind-the-scenes, everyday tales that go on when scientists study the evolution of power.

Caroline Casey, who studies northern elephant seals (*Mirounga angustirostris*), likes to call their social system "a living soap opera." The data suggest that's not hyperbole. The probability that an elephant seal pup will survive to its first birthday is about 35%, and it drops to about 16% for survival to age four. But it's an especially hard-knock life for a male elephant seal. Should he be lucky enough to make it to age six or so, each breeding season he'll engage in a never-ending series of contests with other males for mating opportunities. Odds are he will lose, as 95% of all males will never sire a single pup, despite living a decade or more. On the slim chance, however, that a male emerges as one of a small handful of dominant individuals, the reproductive bounty is great.[6]

Casey worked with Burney Le Boeuf, who has been investigating power in elephant seals for a very long time. He began in 1967, and from the start, the seals' behavior was not at all what the literature he was reading had led him to expect. At that time, research was dominated by studies of power that pitted two individual animals against each other in a lab setting. If the pairwise matchups involved enough different individuals, researchers would sometimes use mathematical techniques to reconstruct a putative group dominance hierarchy. But when Le Boeuf saw elephant seals, he saw a fully functioning dominance hierarchy in nature. The wheels began spinning, and over fifty years later, he's still studying power in elephant seals.

Año Nuevo State Park, just up the road from Santa Cruz, California, is home to the elephant seal rookery that Le Boeuf, his students, and his colleagues have been studying for the last five decades. For the first few years they worked there, the seals were living on one of the islands within Año Nuevo's boundaries. Coordinating boat rides every time Le Boeuf wanted to watch the animals, which was just about every day, became a logistical nightmare. Then he caught a break. In the mid-1970s, with that island shrinking and the elephant seal population growing, a large number of the animals started a new colony on the park mainland, along stretches of coast abutting sand dunes. The dunes, which led to more inland wetland marshes and coastal vegetation, made a perfect home for the seals, and for Le Boeuf, it meant he could leave campus and be at his study site in twenty-five minutes.[7]

Both male and female elephant seals spend months at sea, diving thousands of feet below the surface to feed on ratfish, dogfish, eels, rockfish, and squid, beefing up for their stints on the sand dunes, where they eat nothing at all. But the year plays out in a different manner for males and females. Each spring, females, who weigh about 1,500 pounds and are about 10 feet long, alight on the coast of Año Nuevo for their annual monthlong molt, in which they shed their hair as well as the upper layer of their skin. Males also come ashore for a month to molt, but they do so in summer. During their molt, males look more like gentle-giant plush toys than power-hungry gladiators. They lie side by side, soaking up the sun, and act about as peaceful as any 13-foot-long, 4,000-pound animal might reasonably be expected to act.

December marks the start of the breeding season and the quest for power among those suddenly no-longer-gentle giants. The males come to the dunes at Año Nuevo first, about two to three weeks before females arrive. A free-for-all occurs then, as about 150 males, their giant trunk-like proboscises flopping in the air as they crawl along the beach on their bellies, work out a dominance hierarchy. The first stage in a contest between two males is an exchange of ritualized displays in which a male presents himself in an upright position and emits a call created, in part, by inflating his proboscis. That call is unique in its combination of tempo and timbre. At the same time he calls, he's slamming his body against the sand so violently that Le Boeuf says it feels like a truck is driving by: "The earth kind of feels like it shudders a bit." But no all-purpose information about the caller's size and/or potential fighting ability per se is conveyed by these calls. When Caroline Casey taped the calls of larger and smaller males in one population and played them to males in another, she found that male behavior did not change according to which calls they heard, as you might expect if the calls encoded bits of information on size or anything related to it. Instead, males remember the vocalizations of *specific* individuals they have interacted with. Those vocalizations act as a tag: a sort of name that can be linked to the actual behavior of an opponent.[8]

Seventy-five percent of all elephant seal contests end with one male skulking away after these ritualized displays, cementing the power relations between that pair. If a contest continues, males charge each other while calling, repeating this over and over until they are so close to each other that when slamming their bodies down, they shoot sand up in the face of their opponent. If the contest is still not settled at that point, which happens only about 10% of the time, violent biting, sometimes resulting in significant bleeding, follows. In the end, one male always leaves the loser. Each contestant then pairs off against another putative opponent, and the process goes on like this over and over for two to three weeks. Most power relations between males have been settled at that point, when females start to arrive.

Pregnancy in elephant seals lasts about eight and a half months, but females don't implant developing young into the uterine wall for three and a half months after an egg is fertilized, which keeps parturition (giving birth) on a neat yearly cycle. When females crawl onto the sand dunes for their four-to-six-week stay starting in mid- to late December, they are within days of giving birth to a single pup, the product of the prior year's mating. When they arrive at Año Nuevo, they cluster in groups on the sand dunes: if they tried to go it alone, Le Beouf says, they would be constantly harassed by males and would rarely be able to wean a pup successfully. The most powerful males, the winners of the umpteen battles that occurred before the females arrived on the beach, each guard one of these groups from other males.

Once a female in a group has given birth to and weaned a pup and is ready to mate again—about three or four weeks after having come ashore—the male guarding that group does everything he can to make sure he is the one to inseminate her before she heads back to sea. A stretch of beach might have ten or more groups, each containing dozens of females, so the benefits of being one of those top-ranked males guarding a group are great.

By the height of the new mating season in late January, life on the coast of Año Nuevo is quite the spectacle. Though most females arrive just a few weeks after the males, new females are constantly coming ashore through mid-December and early January, so at any given time some females are pregnant, others have given birth and are nursing pups—who can gain 10 pounds in a single day—and still others have weaned their pups and are receptive to mating. The vast majority of the males who are not top-ranked and not guarding a group are constantly trying to make their way into one group or another to mate with any receptive females. A male guarding a group sometimes staves off such attempts at usurping his power by calling loudly, which works when the usurper recognizes that male's call. But fights sometimes occur if the two have never interacted before. As if all that chaos was not great enough, even this late into the breeding season, a new male occasionally comes ashore, and at that point, there will be a series of contests between the newcomer and the resident males to work out the newcomer's position in the power structure.

Through all the tumult, consistent patterns emerge. The four most powerful males, guarding the largest groups of females, typically get 80–90% of the matings on a stretch of the sand dunes.[9] Those percentages are impressive, but the absolute numbers are even more so. Top-ranked males guard groups with up to a hundred females. Le Boeuf enjoys telling the story of a male that was at the top of the power structure for four years in a row, a rare event given the energetic demands of maintaining that rank year after year. He estimates that male mated with 250 females.

Whether it's percentages or absolute numbers, one assumption when working with elephant seals is that matings translate into successful fertilizations. In most species, that assumption could, at least in principle, be tested by matching molecular genetic tests of paternity against observed matings. But in the mid- to late nineteenth century, northern elephant seals were hunted to the brink of extinction, perhaps down to a few dozen individuals, for their blubber. Even though modern populations have largely recovered from near-extinction numbers, the "genetic bottleneck" they went through means that their genetic diversity is very low—too low to allow paternity to be assigned by genetic testing. So Le Boeuf and his colleagues don't know for certain that insemination always leads to fertilization, but they think the odds are good that it does.[10]

Females are hardly passive players in all this drama. For try as he might, if the group that a powerful male is guarding is sufficiently large, he simply cannot stop all intruder males from entering. When an incursion is successful, the intruder will quickly attempt to mate with a female (and often many females) before he's detected. Oftentimes, the females appear to want no part of such matings with lower-ranking males, and they make that abundantly clear to the intruder and, perhaps more importantly, to the male guarding their group. As the intruder male goes through the typical motions of grasping the female and pinning her down to mate, she will swing her hips back and forth to dislodge him, and use her flippers to kick sand into his face, while emitting a piercing "croaking" sound that Le Boeuf suggests conveys "to everyone in the surrounding area . . . that a female is being mounted." The lower an intruder male is on the power totem pole, the more likely a female is to emit such a call. Upon hearing that call, the dominant male rushes over, dislodges the intruder, and chases him from the area, often following that up by mounting and mating with that female himself.[11]

What of the mating success of these powerless, low-ranking individuals that make up most of the males on the beach? If incursions into a group of females typically fail because dominant males are on guard, and females pick up the slack should an intruder slip under the radar, what options remain? There's only one: after a female has mated with a dominant male, she has to get to the sea, which can be anywhere between 3 and 50 meters from where her groups is. On her way, she may be harassed by as many as twenty low-ranking males attempting to mate with her. "A female has to run this gauntlet of males in order to get to the water and go back to sea and forage," as Le Boeuf tells it. A female can mate after she has been inseminated by a dominant male, though she clearly tries to avoid doing so, as evidenced by the fact that she will use the most direct route possible to the sea and prefers leaving her group at high tide when the distance to the water is at its lowest. But usually that's not enough, and one or more males will attempt to copulate with her. After living off body fat for a month, giving birth, and weaning a pup, a female may have lost 40% of her body weight while at Año Nuevo, and her ability to fend off such attempted matings is low. "It is a dangerous situation," Le Boeuf notes, explaining that males bite females' necks during copulation and may accidentally pierce the large veins that run down the females' spines. If that happens, Le Boeuf says, "she is dead on the spot." As a result, many times females do not aggressively resist such mating attempts and copulate with one of these males while en route to the water. What's unclear, because the assigning of paternity through genetic analysis is impossible, is what proportion of these matings translate into fertilizations.[12]

Once the females have made it back to sea, the males soon head out, too, and the sand dunes then house only weaned pups, who remain for another month or so before heading out to sea themselves. The following December, the whole cycle, including all the power machinations, begins anew. And, as ever, there will be nothing subtle about any of it.

For subtlety, one place to turn is to Whyalla, on the coast of southern Australia, where 185,000 masters of disguise struggle for power.

Film and television producers are constantly bombarding Roger Hanlon with requests to air the underwater videos that he has taken at Whyalla Bay, about 250 miles northeast of Adelaide. If you sat down to watch one of these videos, you'd see a fairly mundane backdrop filled with sea grass, mud, sand, and nondescript rocks, and might think, "What's all the fuss about?" Until one of the 2-foot-long rocks moved, then swam away at full speed, squirting an ink jet as it did. For *Sepia apama*, the giant Australian cuttlefish that Hanlon has studied for over twenty years, is a master quick-change artist.

Hanlon and his team have found that these cuttlefish use many different disguises, each fine-tuned to the background against which they blend. If their background is composed of solid, blackish-gray rocks, giant cuttlefish display a uniform camouflage pattern. More often, they display a mottled camouflage pattern, in which they produce small dark and light splotches that mimic the patchy nature of most of their underwater seascape of grayish small rocks that are sometimes textured with spots of dark algae. On occasion, cuttlefish will even turn on disruptive camouflage, in which they produce alternating large light and dark stripes, visually breaking up (disrupting) their bodies, and creating the appearance of something that looks nothing like a respectable *Sepia apama*.

Even more remarkable than the variety of these disguises is the speed at which giant cuttlefish can camouflage themselves: they can seamlessly blend into the background in a matter of a second or two. Cuttlefish have unusually good night vision, as do their predators, and Hanlon, employing a small remotely operated vehicle with a camera and a red-light filter, has filmed cuttlefish performing their camouflage magic in what our eyes (but not theirs or their predators') perceive as almost total darkness.[13]

Much remains to be learned about exactly how they do any of this, but this much is known: in cephalopods—which include octopuses, squid, and cuttlefish—chromatophores house the pigments used for pattern and coloration. These chromatophores operate differently in cephalopods than in other animals. Rather than being cells under the control of hormones, as they often are in other species, cephalopod chromatophores are sacs of pigment controlled by muscles, which makes them an integral part of the neuromuscular system. The muscles around chromatophores are controlled by various lobes in the brain, primarily those associated with vision. When signals that excite the muscles around chromatophores are sent from the brain, these muscles contract, expanding the chromatophores; when the muscles relax, the chromatophores contract.

These expansions and contractions allow the animals to change not only color and tone, but also skin pattern. Behavioral work by Hanlon and others has found that the way that chromatophore activity is guided by brain signals involves integrating information on background pattern, intensity, and contrast as well as polarity, depth, and the three-dimensional nature of objects.

How all this information is coded and shuttled through the brain to the muscles around chromatophores to produce a desired outcome is still being worked out.[14] Regardless of the detailed mechanics, once that ability evolved—probably as a means to escape detection by seals, dolphins, and other predators—it was co-opted for use in other contexts, including aggression and the quest for power.

Since the late 1990s, Hanlon has taped hundreds of hours of cuttlefish life at Whyalla Bay, where every year, 185,000 giant cuttlefish living in Spencer Gulf make their way to a 4- to 6-kilometer stretch along the coastline to mate in water about 5–20 feet deep. During this giant cuttlefish orgy, females may mate seventeen times and lay between five and forty eggs every day, and they could not care less that Hanlon is watching. "There is no habituation required," he emphasizes. "These animals, they are there to spawn. You approach them slowly and kind of kneel on the bottom and you start filming. . . . It's just unbelievable. You're just like a rock there and you see everything."

The central power players at the spawning grounds are what Hanlon describes as "consort" males. Each of these large males guards a female as she lays her eggs under the face of a table-sized rock, often hidden from view in an underwater forest of undulating plants. "Most of the females have a consort male with them," Hanlon notes. "Occasionally you will see that pair somewhere not being bothered, but what you really see more often, is the female down a little depression and the consort male, which is about double her size, hovering over her." Trouble for these consort males lurks everywhere. The sex ratio is heavily male biased, and as with the elephant seals, guarding females requires constant vigilance. But unlike intruding seals, cuttlefish intruders come in two very different forms.[15]

Smaller males are everywhere, poking and probing. "When you

lock your eyes on one pair," as Hanlon describes it, "you see four or five smaller males hovering around doing different 'sneaker' tactics. It is really dynamic . . . one sneaker will dive in and try to get to a female and the consort male will push him away and then you will see another one try it, and the consort is busy like crazy." A smaller male will almost always flee the scene once a consort male detects and approaches him. But there is a subclass of these small males that have found a creative way around the defenses of the powerful consort males: they disguise themselves as females. Using the ability to change their color and pattern at will, they mimic the mottled pattern typically seen in females. But because males have four arms, but females have only three, mimics are still flashing a telltale sign of their deception. So they retract that fourth arm, hold the remaining visible arms in a posture like that of egg-laying females, and often slip in under the consort male's radar. Hanlon watches (and tapes) these mimics in real time, and DNA analysis has found that they don't just circumvent a consort male's normal defensive behaviors, but sire some of the offspring of females ostensibly under a consort male's guard. Sometimes brain outwits brawn.[16]

It's another matter altogether when it comes to the threats consorts face from a group that Hanlon has dubbed "lone large males." These males will challenge consorts in ways smaller males will not, and because they may be as large and strong as the consort males, simply making it clear that they have been detected will generally not get them to leave. This brazenness allows Hanlon to record them on videotape when they challenge the consort. "They just come up and fight," he says, "and you can film away." Analyses of these tapes, as well as others made in follow-up laboratory work with Alexandra Schnell at Macquarie University in Sydney, shows that these encounters, which range from thirty seconds to twenty minutes in length, include changes in color/pattern, behavioral displays, and full-fledged aggressive interactions. Encounters often escalate through various stages, each of which involves mutual assessment of each other by the contestants.

The first stage of a contest usually has one contestant presenting a "frontal display" to the other. Here, a male faces his opponent and holds his mantle down, while his white-colored arms are oscillating slowly to and fro. His opponent will typically respond in kind, but sometimes retreats. Sans a retreat, stage 2 of the contest involves "lateral displays" and "shovel displays." In a lateral display, a male extends his arms and body sideways, with his fourth arm expanded. During this display, the male will often show a "passing cloud" pattern on his mantle. To create that pattern, he expands and contracts his chromatophores, creating changes in contrast and flow in a way that produces the appearance of undulating, unidirectional dark and light bands. An opponent typically responds with his own lateral display, but on occasion will retreat. Shovel displays are similar to frontal displays except that the mantle is visible and the arms are held in a shovel-like shape.

If a shovel display does not lead an opponent to retreat, then either lateral pushing or more aggressive "frontal pushing" follows. If a contest reaches this third stage, it most often involves lateral pushing, with the two cuttlefish shoving one another back and forth until one retreats. Each stage along the way provides more and more information about an opponent's size, motivation, and strength relative to one's own. In rare cases, when the situation is still not resolved after stage 3, things can get really nasty. A full-fledged fight unfolds, with contestants twisting and turning in three dimensions and, as Hanlon describes it, "leaping over each other biting each other and ink going all over the place." One contestant always retreats as a consequence of such a fight.[17]

At any stage of a contest, if all things are equal and the consort male and lone male intruder are about the same size, the lone male typically retreats and the consort male retains power. Of course, all things may not be equal, and if the lone male is considerably larger than a consort, he may win the contest and assume the role of consort himself. And all else may not be equal in another, less obvious, way. For when Hanlon, Schnell, and their colleagues studied the micro-details of fights, they found that 60% of male cuttlefish favor their left eye when fighting, 25% show a right-eye bias, and about 15% show no bias toward either eye. And, though males with a left-eye bias show a tendency to escalate fights more often than other males, males with a right-eye bias are more likely to win contests. Nothing is simple when it comes to cuttlefish power.

Sometimes the path to power requires not explicit confrontation, but rather espionage, as seen in the green swordtail (*Xiphophorus helleri*). The brain of this fish can comfortably rest on the head of a pin, but when Ryan Earley joined my lab to start his PhD in 1997, he had no intention of letting that stop him from studying power in this fish. And for good reason, too. Earley had been involved in some animal behavior work as an undergraduate at Syracuse University, and he understood the literature well enough to realize that a lot of baseline information about dominance, aggression, and power in swordtails was already available.

By the time Earley started graduate school, animal behaviorists had been running laboratory experiments on aggression in male swordtails for almost thirty years. Size really matters in power struggles among these fish. As a general rule, if a male is more than 10% larger than its opponent, he wins a fight. A number of studies on the hormonal underpinnings of power in this system had also found that males with higher baseline levels of androgens, such as testosterone, attacked and bit opponents more often than males with lower baseline levels.

Put two males—each about 2.5 inches long—together in an aquarium, and what follows is a series of chases, with both fish displaying erect dorsal fins. Next, they'll start nipping and flashing lateral displays, twisting their bodies into an *S* shape. Then each fish straightens out his body and rams it into his opponent, and both fish beat their tails against each other. If that doesn't settle things, they circle each other rapidly, lock jaws, and mouth wrestle, thrashing about until a clear victor emerges. That fish then swims freely about the aquarium, occasionally reinforcing his power status with a quick chase of the subordinate, who typically signals that role by folding in his dorsal fin, retreating, and hovering in a corner or at the bottom of the tank.[18]

In the early 1990s, Dierk Franck, who had led much of the earlier laboratory work, went down to Lake Catemaco in Veracruz, Mexico, to see just how all these controlled studies of fighting mapped onto power in the wild. For three weeks, Franck sat on the banks of various creeks flowing off the lake, using his 7 × 20 binoculars to watch groups of eight to ten male swordtails swimming in water about a foot deep. After a bit of practice, he could recognize individual males by their size, color, and sword length. Based largely on the attacks and retreats he saw, Franck categorized the males in a group as dominant (more chases than retreats), middle-ranked (about the same number of chases and retreats), or subordinate (more retreats than chases).[19]

Though Franck did not see the hyperaggressive mouth wrestles he had observed in the lab, the swordtails of Lake Catemaco displayed the same sequence of aggressive behaviors that those more controlled studies, already firmly ensconced in the ethological literature, had found. It was reassuring to Franck's team that the field observations suggested that swordtail aggression in the lab was not an artifact, and it was also reassuring to Earley, because from very early on, he had been thinking about looking at very subtle aspects of power among swordtails, which, by necessity, would have to be in done in the lab. "It was very clear, just from watching them, there was a power struggle, that every individual wanted to be at the top," he says. "What got me interested was 'How do you get there?' and then 'If you get there, then how do you stay there?'"

Earley first set his sights on answering a relatively unaddressed question about the dynamics of power. Swordtails, like so many other animals, form what ethologists call a linear hierarchy, in which the top-ranked (alpha) individual wins the majority of its interactions with each and every member of the hierarchy, the second-ranked (beta) individual wins the majority of interactions with everyone except the alpha individual, and so on down the hierarchy. In the short term, these hierarchies are stable. But in the medium term, groups dissolve or merge with other groups, and fairly little is known about the redistribution of power when that happens. And so Earley decided to experimentally merge groups with already established hierarchies to see how power transferred (or failed to do so) upon the formation of the new larger group.[20]

To keep things as simple and as tractable as possible, the starting groups in the experiment had three male swordtails housed in their own little aquaria. Earley would sit behind a black curtain that had a slit through which he could watch a group. He'd choose one fish and watch it for fifteen minutes, noting everything it did, as well as every behavior directed at it, on a pad of paper in his lap: every erect dorsal fin, nip, lateral display, or tail beat, and every mouth-wrestling bout the focal male was involved in, including whether it retreated or whether another fish skedaddled from it in defeat. He then cycled through group members, over and over, and did this, day after day, until a clear linear dominance hierarchy emerged. Next, he took two groups in which the hierarchy had been established and merged them into one group by placing all of them in a single tank, and started the observation process all over again, watching the fish until a stable hierarchy emerged, typically within four days, in the new larger group.

After more than four hundred hours of pre- and post-merger observations, one thing was clear: power transferred cleanly when groups merged. The dominant individual in each group of three almost always held one of the top two ranks in the merged group of six, and the middle-ranked and lowest-ranked fish showed a similar pattern. Sitting behind that black plastic screen was a "prime place to incubate ideas," Earley remembers, "because you are just stuck in this black enclosure, watching the fish. It's just you and the fish . . . and then it just popped out at me. These fish weren't just automatons swimming around and moving around their environment. . . . It was about who was watching who and what sort of information was floating around that social environment." They were, it seemed to him, spying on one another. And that called for another experiment.

Earley was in relatively uncharted waters with this new spies-in-the-water eavesdropping idea. Eavesdropper effects, also called bystander effects, occur when the observer of an aggressive interaction changes its assessment of the fighting abilities of those it has observed. The information is available and free, if you have the smarts to process it, but ethologists were just starting to think about this sort of thing when Earley was working with swordtails, and he recalls that "it was thrilling to try and come up with this experimental design . . . sitting down with paper, just mapping out all sorts of different treatments to figure out if they were watching each other and using this information." But he struggled to come up with a design that would allow bystanders to see fighters, but not vice versa, until he experienced "one of those Eureka moments," sitting, in all places, in a car parts shop: "[I'm] getting my nasty old Toyota fixed, and I see these folks who work with limousine tint. And I ask one of them, 'Would that limousine tint cause a reflection?' and he says, 'No, that's the whole point.' That's when I bought a ton of limousine tint, tinted the one-way glass I had, and was able to have it so the bystanders could see the fighters, but not the reverse."

He immediately set up an experiment in which a spy fish swam freely on one side of an experimental tank and a pair of swordtails that were involved in aggressive interactions were on the other side, separated by the tinted glass so that the spy could see in, but the fighters could not see out. Earley also ran a control treatment identical to the first, except that the solitary fish could not see the fighting pair because an opaque partition was placed between it and them. Next, the spy—or, in the control treatment, the solitary fish that could not see what was happening across the way—was pitted against either the winner of the observed fight or the loser of the observed fight.

Spies were much more likely to avoid the winner of a contest they had observed than were fish in the control treatment, which makes perfect sense, and was just what a good spy should do. But what really struck Earley was that while spies were quite aggressive when interacting with losers they had seen quickly retreating from other fish, they were much more cautious in their subsequent encounters with fish who had put up a good fight before losing. What's more, in a follow-up experiment, he showed that watching a fight did not change how a spy behaved toward just *any* swordtail. When spies could see a fight, but were then paired against a naive fish that had not been part of that fight (or any fight), they did not treat the naive fish any differently than would an individual who had no opportunity to spy.[21]

Earley has moved on to other questions and other species, but work on eavesdropping in fish continues, including studies looking inside those small, but apparently powerful, brains of theirs. When João Sollari Lopes, Rodrigo Abril-de-Abreu, and Rui F. Oliveira compared patterns of gene expression—when genes turn on and off and how much protein they produce—in the brains of zebrafish (*Danio rerio*) that could spy on a fighting pair, they found that spying triggered a sequence of complex genetic changes linked to both alertness and memory formation.[22]

Our observations of paths to power in hyenas, northern elephant seals, giant cuttlefish, and swordtails hint at the importance of the costs and benefits of power. Those costs and benefits are fascinating in their own right, but how do they drive natural selection as it acts on the behaviors used by animals to acquire and maintain power?

2   Weigh Costs and Benefits

> I am no more lonely than the loon in the pond that laughs so loud.
> HENRY DAVID THOREAU, *Walden; or, Life in the Woods*

As chief operating officer for the Royal Academy of Arts in London, Tzo Zen Ang spends most of her time these days on terra firma. It used to be different. In a previous life, as part of her PhD work at the University of Cambridge, Ang spent much of 2007–2009 scuba diving in coral reefs around Lizard Island, Australia, watching power play out in dwarf angelfish (*Centropyge bicolor*). She chose her subject of study in part because of its underwater habitat. "I have a love of scuba diving," Ang recalls. "I thought, OK, fish, let's do that."

It was in the crystal-clear lagoons around Lizard Island, at locales like Mermaid Cove, 130 meters long and 30 meters wide, with its giant clam beds, manta rays, and turtles, that Ang first encountered dwarf angelfish swimming 2–13 meters below the water surface, doing what they do most of the day: feeding on algae and detritus in the reef. She'd dive in the morning, in the afternoon, and at dusk, and soon she could identify many an angelfish by sight, relying on a combination of size and color pattern, particularly where the vertical line dividing blue from yellow was located on the body. To facilitate recognizing each and every fish, she captured them and injected them with pink, orange, blue, and green synthetic dyes. In time, she came to know 140 adults, living in thirty-seven different groups, along four different areas on the reef: "Each group has a specific territory," she found, "and they don't wander out of their territory."

Using these groups, Ang began to piece together the costs and benefits of power in angelfish. Each group has a strict linear hierarchy whose order is determined by two factors. At the top of the hierarchy sits the lone male in a group, who is also *always* the largest individual. All of the other adult individuals in a group are females, and their rank is determined, again, by size, with the largest female ranked second in the group, and so on down the hierarchy to the smallest angelfish. And it's all relative. A male will be the largest fish in his group, but the largest female in another group might be larger than he is, yet she is always smaller than the male in her group.

These are not your typical males and females, though, because dwarf angelfish are hermaphrodites, and they display a particular type of hermaphroditism called protogyny, in which a fish is first a female but then, depending on social conditions, may, over the course of weeks, transform into a fully functional male. In terms of power in angelfish groups, the key social factor affecting sex change from female to male is the presence or absence of a dominant male in a group. The only way for a female who holds second rank in a group to rise to the top slot is for the male above her to disappear. Such disappearances are rare, but they happened, on occasion, in the groups Ang tracked: "You would show up one day and the male would be gone," she says. "Who knows [why]? It probably got eaten." At that point, the top-ranked female began to transition into a male, and within a few weeks, rose to the status of the alpha individual in a group.

Within groups, the vast majority of aggressive acts are used to establish and maintain power during bouts of foraging: one fish swims rapidly toward another, displacing it and forcing it to feed elsewhere. On occasion, aggression ramps up to longer, more intense chases, and in rare cases, the aggressor nips its victim. Though the behavioral repertoire associated with aggression is similar in both sexes, the nature of the costs and benefits of power depends on whether an individual is currently a female or a male.[1]

At dusk, during her last dive of the day, Ang would watch the male at the top of the hierarchy amass one of the benefits of being in power. Within a group's territory, each female has a home range, where she spends most of her time. At dusk, the male in the group begins visiting these areas to spawn with the resident female: he shedding sperm, and she eggs. On any given evening, a male may spawn with one or many different females, but each female, if she spawns at all that evening, spawns once, and so the male spawns, on average, many more times than do females in the group. Power pays.

These nightly conjugal visits to female home ranges do come with a price tag. As he moves from one spot to another, the male also patrols the periphery of the group territory to stop dwarf angelfish males from adjacent territories, as well as angelfish from closely related species, from entering. Only the male in a group patrols the territory, and it is an energetically expensive business, not just in terms of the cost of extra swimming, but because about 25% of a male's aggressive acts in a territory occur at the border. What's more, when a male is patrolling (and spawning), he is not foraging, so his average feeding success is less than that of females in the group. And, though there are no hard data one way or the other, Ang thinks these nightly solo patrols lead to higher rates of being attacked and eaten by predators.[2]

Though all females in a group spawn less often than the top-ranked male, one of the benefits of being a high-ranking, powerful female is that you spawn more often than low-ranking females do. It's not exactly clear why males preferentially mate with high-ranking females, but it may be because larger females can produce more or larger eggs. Another perk of power is that high-ranking females have larger home ranges that probably provide more and better food.

There is a subtle, but very real, cost to females for maintaining their positions in the power structure. The male in a group is equally aggressive toward all the females below him in the hierarchy. He is just as likely to get into a scuffle with the highest-ranking (largest) female as with the lowest-ranking (smallest) female. Not so for the females in the group. Not surprisingly, they tend to be aggressive toward those below them, rather than those above them, on the dominance totem pole, but it is more complicated than that. The higher the rank a female holds, the more likely she is to direct aggression to the fish *directly* below her in the hierarchy, who is also the most dangerous (female) competitor she could challenge. Why not just pick on females further down the hierarchy? Ang hypothesizes that it's because the top-ranked female is poised to become the dominant male in a group should the current despot disappear, and so it may be worth the cost of being aggressive to the most dangerous challenger to increase the chances that it does not usurp her high-ranking position. This strategy seems to be effective, as aggression by high-ranking females leads to decreased foraging by those directly below them, which in turn decreases the likelihood of lower-ranking fish moving up in the ranks.[3]

When it comes to the costs and benefits of power, what animal behaviorists want to measure is the net effect—the benefits minus the costs—of different behaviors on long-term reproductive success. Ang's study on angelfish goes a long way in that direction; Holekamp's long-term study on reproductive success in hyenas, which we discussed in the last chapter, sets the gold standard.

Gathering long-term data on reproductive success is easier said than done. It requires continuous monitoring of populations—knowing who's who, which individuals display which different behavioral variants, and how many offspring individuals produce—year in and year out. Every step in that chain has its own logistical problems. As a result, researchers sometimes gather data on a single round of reproduction and use that as a proxy for lifetime reproduction. More often, ethologists rely on second-order proxies, such as the relationship between power and foraging success or between power and access to safe refuge. The assumption is then that more food or better shelter eventually leads to greater lifetime reproductive success. These proxies, and others like them, may also be used to gauge the costs of power. In some cases, researchers can collect information on costs and benefits; in others, only on costs or only on benefits.

For any behavior—including behaviors used to acquire and maintain power—to be favored by the process of natural selection, the benefits must outweigh the costs. In the 1970s and 1980s, to develop a theoretical framework for studying the costs and benefits, and hence the evolution, of power, animal behaviorists borrowed economic game theory from the field of mathematical economics and applied it to nonhumans. Game theory, whose founders eventually won a Nobel Prize for their work, assumes that the payoffs a person gets for an action depend not just on what she does, but also on the behavior of the individual with whom she is interacting. For example, in the classic prisoner's dilemma, if two suspects are apprehended for a crime, the ultimate fate of each suspect depends not just on what she tells the police, but on what her co-conspirator tells the police.

What evolutionarily oriented ethologists did was to take game theory models, make some important tweaks, and use them to understand and predict nonhuman behavior. Evolutionary game theory models were soon being built to predict the best strategy for an animal to acquire power, given how its opponent acts. Game theory is perfect for modeling the dynamics of power because an animal's payoff when contesting a resource depends on whether its potential opponent is willing to fight or backs down when threatened.

A simple game theory model of the evolution of aggression and fighting behavior helps set the stage for how cost-benefit thinking can be used to begin to understand one not-so-subtle aspect of power. In an early attempt to model one component of power, John Maynard Smith and George Price added a multigenerational, evolutionary twist to classic game theory to construct what is called the hawk-dove game. Hawks and doves in this game aren't birds. Instead, the sobriquets come from the political science literature, where aggressive individuals, groups, or countries are dubbed hawkish, while more pacifist ones are called doves. In Maynard Smith and Price's hawk-dove game, hawks use a simple strategy: *always* be ready and willing to fight for a resource. Doves play by a different rule: bluff a readiness to fight, but back down if challenged.

All game theory models are about costs and benefits, and the potential cost to players in the hawk-dove game is the risk of injury during a fight. Let's assume that only the loser of a fight pays that cost, and that the potential benefit is the value of the resource being contested: for example, a food item. Now, imagine two animals contesting a food item. If one uses the hawk strategy and the other uses the dove strategy, there is no fight, since the dove backs down, and the hawk receives the benefit while the dove gets nothing. If both individuals are hawks, there is a fight: one hawk wins and gets the resource, and the other loses and pays the cost. Since all hawks are assumed to be equally good fighters, there is a 50% chance that hawk 1 wins (and gets the food) and a 50% chance that it loses (and pays the cost), so the average payoff to a hawk when it gets into a brawl with another hawk is one-half the benefit minus one-half the cost. Finally, when two doves meet over food, there's some bluffing, but no fighting, and the model assumes they split the resource, with each receiving one-half of the benefit.

After a bit of mathematical analysis, the hawk-dove game predicts that we expect to see *either* a population made up of a mixture of hawks and doves *or* a population composed solely of hawks. Which of the two outcomes we see depends on the exact values of the benefit and cost: increase the benefit and the hawks do better; increase the cost and the doves do. In a population made up of both hawks and doves, power plays include fights between hawks, bluffs and backdowns between hawks and doves, and peaceful interactions between doves: in a population made up solely of hawks, fights and only fights are expected. Perhaps most interesting, though, is that the hawk-dove game *never* predicts a population composed solely of pacifist doves. One way or another, this simplest of models predicts that power struggles should be present.

The hawk-dove game provides a nice example of a thought experiment in which, using only the most abstract notion of cost and benefits, animal behaviorists make some simple predictions. We'll return to the hawk-dove model later when we look at power in cichlid fish, but for now it serves as a nice sortie into a more detailed look at the costs and benefits of power around the planet, including the loose silt environs of the bottom of Michigan's Douglas Lake.

Scattered across the muck of Douglas Lake are ridges made of iron outcroppings. Rusty crayfish (*Orconectes rusticus*) use these outcroppings as shelters, hunkering down there when their predators are active. Contests between these crayfish erupt over these safe havens. To document the arsenal of weapons that crayfish employ when fighting for shelter, and to start to piece together the power structure in *O. rusticus*, Arthur Martin and Paul Moore constructed an elaborate system of underwater crayfish-cams.

What the tapes from these cameras show is an impressive behavioral repertoire: approaches, retreats, tail flips, antennal whips, and more. But it is not until things move past these earlier stages of a contest that crayfish turn to their most dangerous weapon: their powerful claws (chelae). The claws, which can make up an impressive one-third of their 4-inch body length, are used initially to push at an opponent's body or for "boxing," in which two crayfish push their closed claws against each other. If that doesn't settle matters, one crayfish may use an open claw to grab its opponent's body, and should the battle continue past that point, the next stage can be rather unpleasant business, described by Martin and Moore as "unrestrained fighting by grasping and pulling the opponent's claws or appendages."

After poring through a thousand hours of videos, Martin and Moore found that power struggles near shelters were often short, lasting, on average, about 18 seconds. Threat displays, including antennal whips, were by far the most common type of aggressive interactions, but about 20% of encounters involved using claws to push or box with an opponent. Perhaps most telling with respect to power is the role that body size played. When a large shelter owner was challenged by a smaller individual, it retained ownership most of the time; but when the owner of a shelter was smaller than its challenger, it usually lost and was evicted.[4]

The quality of shelters in Lake Douglas was impossible to assess, and so to dig deeper, and determine whether larger, more powerful individuals not only retain shelter ownership more often, but also have preferential access to the best shelters, Martin and Moore brought fifty male rusty crayfish into their lab at Bowling Green University, where they could control the quality of shelters.[5] Ten groups of five crayfish were each placed in their own large aquarium. Each aquarium contained segments of opaque plastic pipe that could be used as a shelter and which varied in size (small vs. large). Analysis of the behavioral interactions within groups found that linear hierarchies were formed. The powerful—those sitting at the top of hierarchy—fought for, and secured access to, the larger shelters. And should one of those larger shelters be occupied by an individual lower in the hierarchy, it was often evicted by a more powerful group member, who then took up residence.

Crayfish show us some of the privileges of power. But it is not all a bed of roses at the top. Costs, as well as benefits, drive the evolution of power.

Parasites, both ectoparasites on the body surface and endoparasites dwelling within, are a nasty fact of life in the wild. But parasites are not equal-opportunity dwellers: some host individuals are mildly infected while others are swarming with both ectoparasites and endoparasites. To better understand the costs and benefits of power, dozens of studies in animal behavior and evolution, including a study on red-fronted lemurs, have looked at how variation in parasite load maps onto power structure.

Under the baobab trees of the Kirindy Forest, on the central-west coast of Madagascar, lives the stuff of fairy tales. Malagasy giant rats (*Hypogeomys antimena*), nocturnal and rarely seen, amaze and terrify humans who are lucky enough to encounter them. Verreaux's sifakas (*Propithecus verreauxi*), better known as dancing lemurs, leap from tree to tree, somehow remaining in an upright posture as they prance about. Dozens of bird, reptile, and amphibian species call the Kirindy Forest home, as do boky-boky (narrow-striped mongooses, *Mungotictis decemlineata*), red-tailed sportive lemurs (*Lepilemur ruficaudatus*), pygmy mouse lemurs (*Microcebus myoxinus*), gray mouse lemurs (*Microcebus murinus*), fat-tailed dwarf lemurs (*Cheirogaleus medius*), and our subjects of interest, red-fronted lemurs (*Eulemur fulvus rufus*).

Peter Kappeler and his colleagues at the German Primate Center in Kirindy have been documenting aggression, submission, and power in these lemurs, who weigh in at about 6 pounds, and whose almost 2-foot-long tails run twice the length of the rest of the body. Each animal has been tagged with a unique nylon collar, and Kappeler and his team record bites and charges, chases and cuffs, grabs and lunges, chutter calls, squeal calls, "huvv" calls, stares, cowers, displacements, stares, tooth gnashing, and "look-aways," to discern winners and losers when red-fronted lemurs fight over food, mates, and other resources.

It might seem like glamorous work studying power in exotic primates on an evolutionarily unique island, but if you want to know about parasites and power, you need to be prepared to collect some feces. In one study that spanned two years, Kappeler and his group collected almost 500 red-fronted lemur fecal samples. The samples were infected with ten endoparasite species: eight species of roundworms, flatworms, and tapeworms and two protozoan blood parasites. Behavioral data on the power status of the sample producers were then mapped onto the parasite information. Though no clear relationship between power status and worm infestation was found, the most powerful males within groups harbored significantly more protozoan blood parasites than other males did. A lemur life of power is also a lemur life with many a blood parasite.[6]

Rather than studying the relationship between power and parasites in a single species, Elizabeth Archie, Bobby Habig, and their team were interested in making large-scale comparisons across dozens of species. They employed a statistical tool called meta-analysis, which takes data from already published studies and searches for patterns. The data for their meta-analyses were extracted from studies indexed on the Web of Science, a massive citation database with information on millions of scientific papers published in more than ten thousand journals. In one of their meta-analyses, Archie and her team focused on male vertebrates and entered two strings of search terms into Web of Science: "parasites, health" and "social status, social hierarchy." They uncovered dozens of hits: studies on rodents, primates, ungulates, birds, ray-finned fishes, and lizards in which data on at least one term in each search string were present.[7]

It turns out that red-fronted lemur males are not alone. Across all vertebrates in the meta-analysis, dominant males had significantly higher levels of both endoparasitic and ectoparasitic infections than did males lower in the power order. For ectoparasites, such as ticks on the fur of impala, part of the reason may be that dominant individuals tend to interact more often, albeit aggressively, with other group members, than do subordinate individuals, and so they may be more susceptible to being infected by mobile ectoparasites that can hop from host to host during contact. The increased susceptibility of dominant individuals to endoparasites may be a result of an inherent, inescapable trade-off: the energy used to attain and maintain power is simply not available for other functions, including fighting off parasites. Still, researchers need to be cautious about inferring cause and effect here. While it seems likely that being powerful causes higher parasite loads (for the reasons just mentioned and more), and rather unlikely that having high parasite loads leads to individuals becoming powerful, what is really needed to nail down cause and effect are experimental manipulations in which randomly selected individuals are parasitized, or not, at a young age, and the effects of this manipulation on the eventual power order are analyzed. In many instances, that's easier said than done. Above and beyond the ethical and legal issues such work in mammals and birds would raise (less so, at least with respect to legal issues, in invertebrates, amphibians, reptiles, and fish), these kinds of experimental manipulations can be logistically very difficult to implement in wild populations.[8]

A high parasite load is not the only health-related cost that the powerful pay, and it's not only primates, and not only males, that pay the cost—powerful female weaverbirds do as well.

As birds go, *Philetairus socius*, weighing in at about one ounce, with its nondescript brown plumage, is not much to look at. But what these weaverbirds lack in looks, they make up for as builders. Their nests—scattered across Namibia, Botswana, and South Africa—are wonders of avian architecture. These communally built structures, some of which have been occupied by *Philetairus socius* populations continually for more than a century, can house hundreds of birds from multiple generations and may very well be the largest, most complex structures built by any bird. It is not without cause that the common name for *Philetairus socius* is the sociable weaver.

Sociable weaver nests buffer the birds from the extreme temperature fluctuations of the African desert. They're constructed of thatched dry *Stipograstis* grass stems, assorted twigs, and materials from the acacia tree on which they usually sit, and with their dome-topped roofs, they can be a yard thick, 3 yards wide, and weigh more than a ton. Deep within, the nest is made up of many, sometimes dozens of, small unconnected chambers, each of which is accessed through a unique entrance from the outside. When a weaver arrives at its entrance, it emits an "entry call," not uttered in any other context, and then proceeds through a tunnel that runs close to a foot long and into its chamber. All chambers are lined with twigs and straws, and each one serves as home to a breeding pair of birds, their eggs, and, on occasion, some extended family members. When not breeding, the pair (and extended family) use the chamber to roost.

As part of a series of long-term experiments at the Benfontein Game Farm in the Northern Cape Province of South Africa, many sociable weavers have been banded with unique number and color combinations. It is extremely difficult to observe behavior *within* nests—though researchers have tried, using mirrors with attached lighting. Fortunately, most behavioral interactions, including those relating to power dynamics, largely play out when the birds are feeding on the ground or in branches around the nest, meeting at one of the external entrances to the nest, or working on the surface of the nest, which they are forever mending and adding new sections to.

The results of clever experiments using artificial feeders placed near nests to facilitate social interactions suggest that power struggles leading to dominance hierarchies are part of the social fabric of *Philetairus socius* life. In one of those experiments, Rita Covas, Liliana Silva, and their colleagues set up cameras near feeders placed beneath each of five sociable weaver nests at Benfontein. For two weeks, starting in late August 2015, they recorded 17,814 interactions between birds, including threats (raising a beak, fluffing out the head feathers), displacements (one bird leaves as soon as another approaches) pecks, kicks, attacks, and more, at the feeders.

They then determined the winner and loser of every filmed interaction in order to piece together dominance hierarchies. Covas's
team also collected blood samples from the birds. They had a special interest in using these samples to measure what's called oxidative damage, which occurs when cells don't produce enough antioxidants, substances that are used to combat certain molecules that
cause damage to the genetic code.[9]

Analysis of the interactions filmed near the feeders found that
sociable weavers form stable dominance hierarchies. Silva's team
then turned to the blood work and searched for any relationship between power within a nest and oxidative damage. What they found
was that powerful females, but not powerful males, paid a physiological cost: higher rank in females was correlated with increased
oxidative damage. Silva and her colleagues suggest that this physiological cost of power in females is somehow linked to acquiring and
then maintaining power. But why, then, don't dominant males, who
are in fact involved in *more* aggressive interactions than dominant
females, pay that same physiological cost? Here, Silva and her team
suggest that foods high in antioxidant properties may be the key.
Because males are dominant to females, they probably have greater
access to foods high in antioxidants, which buffer them from oxidative damage and its consequences.[10]

A caveat is in order here. Parasites and oxidative stress are but
two of many negative indicators of health in animals. Measurements
of other indicators, especially hormones associated with stress, suggest that those lower on the power scale pay the greater price. Subordinate female meerkats certainly do, though that's only one of the
many things that make meerkats especially interesting subjects for
studying power.

Matthew Bell, a behavioral ecologist now at the University of Edinburgh, will be the first to tell you that when it comes to the costs and
benefits of power, what matters is not *absolute* reproductive success,
but *relative* reproductive success: how an animal's reproductive success compares with that of others. That means that sometimes, as in
the case of the meerkats (*Suricata suricatta*) of the Kalahari Desert

that Bell studied, suppressing the reproduction of underlings can be a potent weapon in the arsenal of the powerful. To understand how and why, we need to understand both the benefits and the costs of this reproductive suppression.[11]

Bell's PhD work was not on meerkats, but rather on pup care in banded mongoose (*Mungos mungo*) populations in Uganda. As he gathered those data, he could not help but notice the way high-ranking pregnant banded mongooses tried to suppress reproduction by females lower in the power order. "The larger, more dominant, female would beat the crap out of the smaller ones. Massive disruption," he says. He wrote one paper on that, published in *Proceedings of the Royal Society*, and began dreaming of follow-up, in-depth studies of reproductive suppression and power—if not in banded mongooses, then in another system.

A few years later, Bell got his chance, when he was back in the Kalahari doing postdoctoral work with pied babblers (*Turdoides bicolor*), at a site where meerkats lived as well. He knew from his prior time there, and from lots of work coming out of Tim Clutton-Brock's group at Cambridge (where Bell had received his PhD as part of the Clutton-Brock coterie), that one striking aspect of meerkat social life is the extreme reproductive skew found within groups. Though all adults are capable of reproduction, in each group of meerkats, only the dominant male and female produce offspring: everyone else acts as a "babysitter," helping to raise the pups of those in power, watching over them, and provisioning them with food. Clutton-Brock's group had shown that dominant females maintain their monopoly by actively suppressing reproduction in subordinates.

Dominant females typically reproduce two to four times per year, and if an upstart subordinate becomes pregnant, the dominant female does not take it well. At first, her aggression toward the subordinate is tempered and is limited to aggressively displacing her from a feeding hole or stealing her food. Then it gets noticeably and measurably nastier. Dominants chase pregnant subordinates, and if they catch them, they pin them down and bite them at the base of the tail or neck. "That can escalate to where the whole group joins in and also attacks," Bell says; ". . . it can be relentless, on and on through the day, and eventually the subordinate will just leave the group for a period of days or weeks . . . until the dominant has given birth, and then [she] comes back." Mortality for subordinates after a temporary eviction is high. And those that survive to eventually return to the group have lost weight, have off-the-charts levels of stress hormones called glucocorticoids, and often abort their pregnancies.[12]

Reproductive suppression of subordinates works, but it is not cost-free. It is energetically expensive for a dominant female to engage in intense aggression when she herself is pregnant, and there is always the risk of physical injury inflicted by a subordinate. What a perfect system to study reproductive suppression by the powerful, Bell reasoned. He began talking with Clutton-Brock about those dreams he had held on to since working with banded mongooses, and together they hatched an experiment to find out what happens when subordinate meerkats *don't* attempt to breed; when, as Bell says, "dominants no longer have to invest in beating up their subordinates." Three times a year, for two years, Bell administered a contraceptive hormone, Depo-Provera, to thirty-five subordinate females across six groups of meerkats. Thirty-eight subordinate females in six other groups served as controls and were injected with a similar amount of a saline solution.

When the study began in 2009, the meerkat work had been going on continuously for more than twenty-five years, and though the animals still lived a very wild life, they were also used to having people around them. But that hardly meant that catching seventy-three meerkats three times a year to administer a contraceptive (or saline) was easy. Bell had a canvas pillowcase that he hid behind his back as he approached a meerkat. If he was lucky and the meerkat was more interested in eating its meal than in another human mucking about, Bell would grab the animal by the tail and put it into the pillowcase. Then, with a gas-powered portable anesthetic machine at the ready, he'd stick the meerkat's nose in the mask, knock it out, and administer Depo-Provera (or saline solution). In a few minutes, the meerkat would wake and go straight back into its group.

Every two weeks over the course of his two-year study, Bell and his group gathered data on the behavior of the dominant female in each of the twelve groups. What they discovered from more than a thousand hours of observation was that dominant females attacked the subordinate females treated with Depo-Provera less often than they attacked the subordinates in the control group, and when dominants were pregnant, they were much less likely to evict the Depo-Provera-treated females, who emitted different odors as a result of the contraceptive suppressing their reproduction, and who did not get pregnant.

For the dominant female, the benefits of having subordinates' reproduction suppressed were real and significant. In groups where subordinates received the contraceptive, the dominant female fed more often and gained more weight than in control groups. Pups produced by dominant females in the treatment groups were born heavier and gained more weight during development than did pups born to dominant females in the control groups—which is no small matter, as size at adulthood affects the probability of becoming dominant and is positively correlated with reproductive success. This weight gain occurred not only because dominant females fed their pups more, but also because reproductively suppressed subordinate females, with no pups of their own, provided more food to the pups of the dominant female than did those in control groups.

The gross benefits accrued by dominants when subordinate reproduction is suppressed are clear. But when scientists like Bell aren't suppressing subordinates, it is costly for dominants to do so, in terms of both energy and risk of injury. Whether the *net* benefits to the dominant are positive is unclear because those energy and risk-of-injury costs are very difficult to quantify. Fortunately, other costs of being a powerful meerkat are easier to measure. Kenda Smyth, Tim Clutton-Brock, and others on the meerkat research team found that compared with subordinates, dominant females harbored more endoparasites, including more roundworms and tapeworms, and had weakened innate immunity.[13]

The costs to both dominant and subordinate females are real at Meerkat Manor. But fatalities in the quest for power are rare there. Not so for the common loons that nest on the lakes of Wisconsin.

As the sun rises and the mist atop a lake in Rhinelander, Wisconsin, begins to recede, a pair of common loons (*Gavia immer*), with their piercing orange eyes, black heads, daggerlike beaks, and checker-patterned feathers, serenely move about the lake surface. A male sings a haunting wail call, and a female swims by his side. Soon a "tremolo" call, which seems to waver and tremble in the air like a strange laugh, echoes across the shore. But things are not as peaceful as they appear, because power plays out in rather gory ways in these birds.[14]

After spending the winter in Florida or the Carolinas, these loons return to their territories on hundreds of Wisconsin's 15,000 lakes. A monogamous pair's territory can, and many times does, encompass an entire small lake. Breeding pairs often stay together for many years and return to their home lake each spring. During the breeding season, the male surveys the lake for a good location for a nest, sometimes settling on a stretch of shoreline, but more often choosing a small "island" rising a few feet above the water surface that is relatively safe from mammalian predators like raccoons and skunks that savor loon eggs.[15]

It's a beautiful system, says Walter Piper, who has worked with these birds ever since the early 1990s, if for no other reason than that "even though lakes vary infinitely, at least you and the loons agree that an entire small to medium-sized lake is their territory." That territory is a valuable commodity, and power in the loon system is about controlling it, because only territorial individuals have the opportunity to mate. Power is about place, and so territory holders are sometimes challenged by intruders, called floaters, who lack their own territory. A male floater will attempt to force out and replace the resident male, violently if need be. "It's a constant battle to hold on to a territory with all these other loons coming in," says Piper. If a male floater is successful, the female stays put, and if the floater is a female and displaces the resident female, the male stays, suggesting that fidelity in the system is to a territory and not to a mate.

Piper, Jay Mager, and Charles Walcott have been investigating floater takeover attempts, with a special interest in how male territory holders attempt to fend off would-be usurpers and maintain their reign of power. About half the individuals in a population are floaters, and on any given day between April and August, a territory may be scouted by many different floaters, one at a time. It is hard to know exactly how often these visits involve repeat intrusions by the same individual, because unlike territory holders, who have all been marked using leg bands, only about 40% of floaters have been marked by Piper's team. From the subset of floaters that are marked, it's clear that some do visit the same territory many times. Piper's data on marked floaters suggest they are "reconnoitering . . . trap-lining, visiting here and there, and covering a lot of ground," gleaning information not just on territory owners, but on whether there are chicks present, which they use as an indicator of territory quality.

When a resident male spots a floater, he produces a "yodel" call, which has an opening burst that rises in pitch and is followed by two short notes, repeated over and over. Because lower-frequency yodels are correlated with greater body mass, a floater can quickly get an estimate of the condition of the territorial male. When Piper and his colleagues experimentally altered yodel calls, they found that intruder males not only *can* estimate the body mass of a resident male from these calls, but that they *do*, as they behave more cautiously in response to lower-frequency yodels. Yodels also signal the aggressive motivation of a territory holder: other experimental manipulations found that when the two short notes are repeated more often, a floater again acts in a more cautious manner, behaving as if the territory holder is prepared to fight.

Most visits by floater males are fairly uneventful. The resident territorial male, yodeling, and often with his mate swimming alongside, approaches the floater, and a series of head bows, circle dances, and other stereotypical, but not especially dangerous, behaviors follows. More often than not, the floater flies off less than thirty minutes later, probably to visit another lake. For the territory holder, the costs are negligible, the benefits great, and the power structure remains as it was. But that's not always what happens.

Of the 425 intrusions Piper and his group recorded, 109 involved bouts of aggression that got very intense—so much so that Piper struggled to keep up with it all. "In the excitement of what's going on," he says, "you're just out there in your canoe and trying to figure out what's going on." Generally speaking, low-level aggression involved one bird chasing the other across the water surface until it left the lake. Escalated aggression, which was observed in 25 bouts, involved lunging at the beak of an opponent, often followed by the birds simultaneously grasping each other's heads and beating their wings against each other on the water surface. If things got really heated, one bird might dunk an opponent's head under water for long stretches. Almost half of the 109 bouts of aggression—and so about 10% of all intrusions—led to a change in the power structure, with the resident male evicted from his territory and replaced by the floater.

Sometimes Piper has seen an evicted territory holder escape after losing the first round, but the (former) floater patrols his new area, and often finds the escapee and attacks him relentlessly. What surprised the loon research team was just how costly escalated aggression could be. Within one week of losing power and being evicted from their territories, at least eight, and perhaps as many as sixteen, of the evicted males the team was tracking were dead, with lacerations around their heads and necks, almost certainly as the result of a series of fights they had lost over the course of days. Those that are lucky enough to survive eviction end up on what Piper's team describe as "vacant, unproductive territories nearby."[16]

This sort of fatal fighting is very rare in animals, and when it does occur, it tends to be in short-lived species that reproduce only once. That makes some sense since everything depends on that one bout of reproduction. But loons are long-lived birds, with some reaching their twenty-fifth or even thirtieth year, and they produce many clutches of chicks. Why are they the exception to the rule? Piper and his colleagues are slowly, but surely, piecing together that puzzle.

Victims of deadly encounters tend to be older males who had held high-quality territories that were home to successful breeding attempts in the past. Piper has found that young floaters target such high-quality territories, and that resident males increase their aggressive yodel calls as they get older, suggesting that they place a very high value on their homes. The problem is that males lose body mass as they age, and they do so at an accelerated rate when their territories are productive, as the energy they expend on chick rearing over the years is considerable. Older territory holders yodel more and become more aggressive to floaters as their own tenure on a territory increases, but because yodels provide information not only on aggressive motivation, but also on body mass, such calls also inform floaters that a resident male may be prime for a takeover.

All of these circumstances help explain why fights over high-quality territories occur, and even why such battles can be so intense, but is it really worth it for the resident male to be willing to fight to the death against a younger, stronger floater? That depends on the alternatives. Young floaters will fight hard to usurp a territory, but if signs start to point to a loss, they don't keep on fighting and risk serious injury or death. They simply move on and try somewhere else. The stakes are different for an older male already in place on a productive territory, as he is unlikely to usurp another productive territory should he be evicted, because, as Piper puts it, "the wheels really start to come off" late in life. And so a desperado "fight to the death, if necessary" strategy may be his only viable option.

Power hands a territorial loon riches, but in so doing, sometimes locks him into a do-or-die contest with dangerous upstarts to retain those very riches.[17]

A lot is riding on how animals behave in their quest for power. The wrong choices come saddled with real costs; the right choices may lead to rich rewards, but even those rewards are never cost-free. And so, one important choice any animal seeking power must make over and over is who to challenge and who to avoid challenging. Another is what to do when challenged: Retreat? Fight? And if fight, for how long? Given what's at stake, you might expect that animals would spend considerable time and energy assessing potential opponents.

They do.

3 Assess Thy Rivals

> From the point of view of the moralist, the animal world is
> on about the same level as the gladiator's show.
> T. H. HUXLEY, "The Struggle for Existence: A Programme"

Thomas Henry Huxley was a Victorian scientist to his bones. He was raised in the dog-eat-dog world of nineteenth-century Britain, where one of his admitted occupations was to engage in "an endless series of battles and skirmishes over evolution." On the day before Charles Darwin published *On the Origin of Species*, Huxley wrote to his friend in an attempt to calm Darwin down: "As to the curs which will bark and yelp, you must recollect that some of your friends, at any rate, are endowed with an amount of combativeness which (although you have often and justly rebuked it) may stand you in good stead. I am sharpening up my claws and beak in readiness . . . prepared to go to the stake, if requisite."[1]

Huxley actively sought the role of Darwin's spokesman: indeed, Darwin referred to him as "my good and kind agent for the propagation of the Gospel, i.e. the Devil's gospel," while others simply used a moniker Huxley created for himself: "Darwin's bulldog."[2] In his position as Darwin's prolocutor, and from his understanding of natural selection, Huxley had a thing or two to say about power. The sentences following the epigraph above, from "The Struggle for Existence: A Programme," state his view of it in no uncertain terms:

The creatures are fairly well treated, and set to fight; whereby the strongest, the swiftest and the cunningest live to fight another day. The spectator has no need to turn his thumb down, as no quarter is given . . . the weakest and the stupidest went to the wall, while the toughest and the shrewdest, those who were best fitted to cope with their circumstances, but not the best in any other way, survived. Life was a continuous free fight, and beyond the limited and temporary relations of the family, the Hobbesian war of each against all was the normal state of existence.

Common loons, and a few other species, sometimes meet the gladiator criteria, but in most cases, the struggle for power is far more complex and interesting than that painted by Darwin's bull-dog, and it requires animals to constantly assess other individuals as well their environment. This sort of nuanced assessment behavior is a potentially potent tool in the struggle to attain and maintain a hold on power. In the frigid winters of Canada, caribou find the time and energy to assess rivals for power, as do hermit crabs on the beaches of Ireland, South American cichlids, bowl and doily spiders in New England, wasps in New Orleans, and skylarks in the South of France.

Looking in on caribou guarding snow craters in Canada is a good place to get started.

When the temperature dropped to −49°C, the woodland caribou (*Rangifer tarandus*), also known as reindeer, that Cyrille Barrette was studying at Grands-Jardins National Park, in the Lac-Pikauba Unorganized Territory of Canada, "would stand and do nothing." But when the temperature warmed a bit, to a balmy −20°C or so, power plays among the caribou heated up with it.

Along its 30 kilometers of trails, Grands-Jardins—embedded in a mixture of deciduous forest, boreal forest, taiga, and tundra— is home to bears, moose, foxes, porcupines, wolves, lynx, Canada grouse, and more. Spectacular views of it all, as well as of the Charlevoix meteor crater 25 kilometers east, can be had from the peaks of Mont du Lac-des-Cygnes (Swan Lake Mountain), Mont du Gros-Bras (Big Arms Mountain), and Mont de l'Ours (Bear Mountain).

The caribou in Grands-Jardins today are relative newcomers. A hundred years back, the area swarmed with them, but intense hunting brought caribou to near extinction across Canada. In the late 1960s, when a large-scale reintroduction program began, a herd was placed in Grands-Jardins. When funding to study this population became available in the late 1970s, Barrette seized the opportunity. By the time he began his work there, the majority of the 150 caribou in the park had been wild-born in Grands-Jardins.[3]

Barrette already knew a thing or two about working with this sort of animal, having done his PhD on barking deer (*Muntiacus muntjac*). From that work, he was optimistic that caribou might provide some insight into power in male deer, as well as the woefully understudied matter of power dynamics in female deer, because caribou are the only species in the deer family (Cervidae) in which females regularly grow antlers.[4]

In late fall 1980, and then again a year later, Barrette and his colleague Denis Vandal headed out from Laval University in Quebec and made the 150-kilometer journey to Grands-Jardins, driving north along Route 138 and hugging the Saint Lawrence River to their east. At Grands-Jardins, they settled into one of the cabins in the park, their home until late April. Just getting from the cabin to the caribou each day was an ordeal. During the winter, they would make the daily one-hour trek on Skidoo snowmobiles to the open areas where the caribou roamed. Barrette and Vandal could fit comfortably on a single Skidoo, but "we always went out with two snowmobiles," he recalls, "for safety reasons. If you get stuck 15 kilometers from the cabin with this kind of weather, even if you have snowshoes, you are in *big* trouble."

Once they found the herd they were studying, they would stand out in the cold all day (there were no blinds or shelters in that remote area), speaking their observations of all things caribou into their handheld tape recorders. They quickly came to know the identities of all the herd members. During the winter, the caribou in Grands-Jardins feed primarily on ground lichens. This poses a problem for the caribou because the lichens typically lie buried under the snow. At the same time, for an ethologist studying power, it makes things a lot easier. "Most of the time they were just looking for food," says Barrette, and "looking for food in the winter in one meter and a half of snow when you feed on ground lichens is a lot of work. They would spend most of their time digging craters in the snow to get access to lichens (or they'd just rest). It made a perfect situation for social interactions . . . animals were constantly displacing one another from craters."

Early on, as Barrette and Vandal took detailed notes at these snow-crater caribou power hubs, they were struck by the impression that male caribou, at least as far as they could tell, were assessing the size of their antlers relative to their opponent's and acting on that information in contests. After 436 hours of watching and recording power struggles, their intuition proved correct: when animals sparred, they were indeed assessing their rivals. All-out fights were very rare (six instances), but sparring bouts were a daily occurrence. They observed 11,640 social interactions, 3,500 of which occurred at snow craters; of those, about 37% involved two males who engaged in a sparring match. Those bouts began with the two caribou positioning and repositioning their antlers, then pushing and twisting. When contact between rivals' antlers ended, neither animal ever charged the other (as it would in a fight). Eventually, the contest ended when one caribou backed away and left the contested snow crater.[5]

If sparring itself was part of an assessment process, and if animals didn't have any information about their relative size and fighting potential *beforehand*, then, Barrette and Vandal predicted, half of the sparring bouts would be initiated by the smaller caribou and half by the larger one—and that's just what they found. Also in line with the assessment hypothesis, the smaller male terminated—and so, in a sense, lost—90% of all sparring bouts, often after just thirty seconds of sparring. But size isn't everything: if a male was much younger than its sparring partner, the younger caribou almost always lost, even if its antlers were larger, suggesting that other things besides antler size were being assessed, although it is not clear exactly what or how.[6]

Barrette and Vandal were also keenly interested in sparring bouts that involved females. How and why caribou are the only deer in which females have evolved antlers are complicated questions. Researchers sequenced the caribou genome in 2017, and a comparison of that genome with those of other deer species provided a hint on the *how* question, which appears to center on hormones. A mutation associated with a single gene produces an extra binding site at a receptor in caribou that allows for low levels of hormones that facilitate antler growth in females.[7]

Why natural selection has favored female caribou antlers is all about power. Female-female interactions are relatively rare, but Barrette and Vandal collected data on 110 sparring bouts between males and females at snow craters. As a rule, these bouts did not go well for the female, regardless of whether a male approached her snow crater, or she approached his: after the sparring was over, males kept, or won new access to, the contested snow crater 84% of the time. But all was not lost for those antlered females, for sparring bouts, by definition, involve two animals with antlers. Males shed their antlers in late December–early January, but females hold on to their antlers until they give birth in early June. Interactions over snow craters between a female and a male, when she has antlers and he does not, always go her way: "As soon as a male loses his antlers," Barrette makes clear, "he loses his social status."[8]

It is easy to fall into the trap of thinking that large, furry creatures with big brains may be capable of doing this sort of complex assessment of the power landscape, but that this kind of strategic behavior is not to be expected in invertebrates. That would be a mistake. You don't need to be a hulking caribou at a snow crater to assess opponents and the environment. It works perfectly well if you are a hermit crab in a tidal pool.

With the possible exception of finding a mate, nothing matters more to the hermit crab *Pagurus bernhardus* than finding and moving into a good shell. As in most species of hermit crabs, the head and the thorax of *Pagurus bernhardus* are calcified, but the abdomen is not. It is always soft-bodied, and that's bad news in terms of protection against predators. Hermit crabs act accordingly: "You never see one without a shell," notes Robert Elwood, who has studied these crabs since the late 1970s. "They just wouldn't survive in nature without

[it]." Shells not only provide protection from predators, but also buffer the crab against changes in salinity and help prevent desiccation.

But not any shell will do. Some shells are better than others. If a shell's too small, it doesn't provide ample room to withdraw from danger; if it's too large, it can be energetically costly to lug around. On occasion, an empty gastropod shell, from either a deceased prior occupant or one that outgrew its home, may be found to move into, and in rare instances a hermit crab may attack an injured gastropod, kill it, and move into its shell. But most good shells are found perched on the backs of other hermit crabs, who want to keep those shells right where they are. It doesn't matter if a good shell is empty or on the back of another crab—if it is a better shell than what a crab currently has, what matters is moving into it, because a good fit leads to faster growth and increased reproductive success. So it's hardly surprising that power revolves around these precious homes-on-a-back.[9]

Elwood's initial exposure to *Pagurus* hermit crabs was a small project that he did as an undergraduate at the University of Reading in the late 1970s. He was fascinated by the way crabs fought over shells, but once the class was over, so, he thought, was his time with the hermit crabs. But in the late 1970s, hermit crabs returned to Elwood's radar when he landed a faculty position at the University of Belfast, in his native Northern Ireland. He was looking for projects he could do locally: "I made some inquiries," he recalls, and learned that about an hour's drive away, there "was a mile stretch of beach which was just ideal for hermit crabs. . . . I could go and get 100 crabs in an hour's collection."

He'd search about, moving from one shallow tidal pool to another, and it would not be long until he found a group of crabs scampering in the rocks and weeds of a pool, especially when the weather was cool. Then it was just a matter of scooping them up and putting them in a bucket along with some seawater and bringing them back to his new lab at the university. From those first collecting trips, he knew the crabs would be a good system in which to study power, as he could see scores of fights in the tidal pools. And in the collecting buckets. On one mission, when his two-year-old daughter came with him to the beach, he remembers her looking down into one of the buckets and saying, "Stop it, bad crabs," as they fought, oblivious to the moral compass of their young spectator.

Elwood focused his attention on males, because female hermit crabs carry around eggs for long stretches, and he thought that this would make studying power and shell acquisition more complex. He began by trying to assess exactly what male hermit crabs were looking for in a shell. He knew that shells from the flat periwinkle (*Littorina obtusata*) were a particular favorite of the crabs, though they never seemed quite satisfied with the shell they were actually in. Elwood likes to compare a crab choosing a shell to a car enthusiast in a showroom. Even if you just bought a car, you still keep an eye out for a better one: "No matter how many shells you give a hermit crab," he says, "... and you think he's got a pretty good one, they will still go over and investigate a new shell."

Elwood's early work on shell preference didn't involve power and fights between crabs, but was rather a series of choice experiments. He wanted more than a juicy analogy about car shows: he wanted to gauge how crabs assigned value to a shell. Preliminary work let him estimate the optimal shell for a crab of a given size, and he then set up a novel experiment in which he manipulated the quality of an empty shell that was available to a crab (25% or 100% of optimal quality), as well as the shell that the crab was currently housed in (25%, 50%, 75%, or 100% of optimal quality). To get each crab housed in the shell he needed to for the experiment, Elwood removed crabs from the shells they were in—"We tickle their abdomen with a paint brush and they will just jump out and run away," he says with smile—and then put them in the appropriate shell. Next, he blocked the entrance to an empty shell by cementing it closed, reasoning that, compared with other combinations, if the sealed-off shell was better than the shell the crab was in, it would spend more time trying to circumvent Elwood's roadblock and access the high-quality shell, which is just what it did.[10]

What really kicked the work on power into high gear was a short paper by Brian Hazlett that Elwood read, in which Hazlett suggested that hermit crabs negotiated over shells. It turns out there is no negotiating involved, but Elwood and his colleagues carried out a series of experiments that revealed the fascinating power dynamics surrounding shell battles.[11]

These battles include not just approaching and retreating, but grasping, climbing on the back of an opponent's shell, probing into an opponent's shell, "shell rocking," where the attacker grabs hold of his opponent's shell and rocks it back and forth, and shell rapping, where an attacker uses his abdominal muscles and walking legs to rap his shell against an opponent's shell. Elwood watched all this, and for the rapping, used a microphone in the water, sealed in a rubber covering (which happened to be a condom), to measure sound pressure, a measure of the intensity of the raps.

Elwood collected data on everything from the number and length of rap bouts to the duration between raps in a bout, the duration between bouts, and, perhaps most importantly, the intensity of the raps, which a crab might use as a way to gauge the strength of an attacker. What he found was that if an attacker persists and a defender decides to give up, then somehow—and Elwood is still not sure exactly how—the loser's decision to submit is signaled to the attacker. At that point, the attacker grabs the defender, yanks it out of its shell, and throws its vanquished opponent to one side. It then moves into the loser's shell, but initially keeps hold of its original shell, just in case a closer inspection of its new shell suggests that a move would be a mistake, which rarely happens. "Only when [the attacker] makes a final decision," Elwood says, "does it walk away, leaving the naked defender to take [the attacker's old shell]."

In an early experiment on power, published in a paper titled "Shell Wars," Elwood and Barbara Dowds compared the behavior of pairs of hermit crabs in four different situations. Crabs were either large or small, and were placed in either a preferred shell from a flat periwinkle or a suboptimal shell from a small sea snail. In two of the treatments in this experiment—when each hermit crab was housed in a periwinkle shell or when each was housed in a sea snail shell— the larger crab had nothing to gain by evicting the other crab from its shell. In the more interesting treatments, pairs were made up of a larger individual in a preferred periwinkle shell and a smaller crab in the shell of a sea snail, or a smaller individual in the better shell and a larger individual in the inferior shell. Using a simple contraption that allowed them to tap different keys on a keyboard to record different behaviors, Elwood and Dowds then acted as correspondents for the ensuing shell wars.

Crabs appeared to be assessing both their relative size (and/or strength) and the quality of their shell relative to others. In 93% of trials, the larger, stronger male initiated attacks, which is a far higher proportion than would be expected if the animals could not assess size and/or strength. What's more, in 75% of the trials when the larger crab was in an inferior shell, it evicted the smaller individual from the preferred shell, securing a better (mobile) home for itself, probably as a result of an assessment of shell type and size and/or strength by the attacker and the defender. The corresponding eviction rate was 3.6% when the smaller individual was in an inferior shell and the larger crab was in a superior shell.[12]

With all the constant grasping, climbing on, probing, rocking, rapping, and evicting going on, shell wars are energetically expensive. Just how expensive is what Elwood and his colleague Mark Briffa set out to explore. For the attacker, the cost was a buildup of lactic acid, a by-product of the metabolic processes used to produce energy: the more shell rapping a crab initiated, the more lactic acid it produced. The results from defenders were not as intuitive: defenders converted *more* muscular glycogen to glucose, and hence energy, when an attacker rapped *less* vigorously. Why, Briffa and Elwood thought, would crabs produce more energy to deal with a weaker opponent? While their data don't directly address that question, they speculate that when a defender senses a weak attacker, it assesses its chances of holding on to its shell as relatively high, and hence worth the energy investment, but if it's clear that an attacker is likely to evict a crab, no matter how much glycogen that crab converts to glucose, better to hold that glycogen in storage for future use.[13]

As Elwood and his colleagues studied power dynamics in hermit crabs, other ethologists were building and testing models of the role of assessment. But none of them thought those models would have anything to say about nuclear war. They were wrong.

"I read about the hawk-dove game, and I remember thinking

this is not how animals fight," says the University of Stockholm's Magnus Enquist. "They have much more information, and I realized they [were using a] sort of statistical process where they keep getting better estimates [over time]." What was needed, he decided, was a new assessment-based model of how power is gauged in animals. Though an avid naturalist from age three, Enquist had chosen to study mathematics as an undergraduate, so he had some mathematical modeling experience. In 1979, as part of his dissertation work at the University of Stockholm, he began building what he saw as the missing model. But he didn't work alone, instead teaming up with Olof Leimar, another PhD student, who at the time was studying theoretical physics. "There were few big problems left in physics and they were very difficult," Enquist notes, and so it was not all that surprising to him that Leimar would be interested in applying the modeling skills he had developed to a question in biology.

When Maynard Smith and Price built the hawk-dove game in 1973, they had imported and modified models from political science and mathematical economics. Mathematical model building in ethology was, relatively speaking, still in its early days when Enquist and Leimar began their collaborative work, so they, too, looked to other fields for tools to build their model. They settled on models from the field of insurance mathematics, which is not as strange as it may sound, because actuaries are quite interested in the statistical sampling process that Enquist and Leimar thought so important in establishing power.

In 1983 Enquist and Leimar rolled out what they called the sequential assessment model. The model assumes that going into a fight, contestants have incomplete information about each other's fighting abilities. As they interact, they constantly reassess their own fighting ability relative to that of their opponent. Assessing an opponent's fighting ability in the model is analogous to a process of statistical sampling. One sample—a single assessment of fighting ability—introduces significant error, but the more samples (assessments), the lower that error rate, leading to more confidence in the animals' relative fighting abilities.

The sequential assessment model examines struggles for power in which the level of aggression varies from mild to dangerous. The model predicts that individuals should begin with the least dangerous type of aggressive behavior, sampling one another until all the available information with respect to *that* behavior is exhausted. Then, the prediction goes, the next most dangerous type of aggressive behavior should be used, and again, after some sampling by the protagonists, all the available information about an opponent and *that* behavior becomes exhausted. More dangerous behaviors are then added and sampled until one contestant decides that its probability of victory is low enough that it should end the contest. Because evenly matched opponents need to do the most sampling to distinguish their relative fighting ability, the model also predicts that such contests should both last the longest and escalate to the most dangerous behaviors.[14]

As it happens, at the same time Enquist and Leimar were building the sequential assessment model, others in Enquist's Department of Zoology at the University of Stockholm were studying the mating behavior of the cichlid *Nannacara anomala*. Native to Suriname, *Nannacara*, whose name translates to "small face," measures about 5 centimeters long from head to tail. Enquist had visited the lab of his colleagues studying *Nannacara anomala* with an eye not so much toward studying mating behavior, but toward studying power. He liked what he saw: males formed dominance hierarchies, and aggressive interactions ranged from fairly innocuous ones, like "changing color" and approaching an opponent, to the more dangerous, including tail beating, mouth wrestling, and "circling," the most dangerous of all, where fish repeatedly attempt to bite each other while they swim in a circular pattern. Aggression comes to a halt when one individual signals submission by folding its fins and changing its color.

Enquist, Leimar, and a team of others videotaped staged encounters between 102 pairs of males. In some pairs, males were about the same size, and in others, one male was bigger. Analysis of the tapes showed that, as the sequential assessment model predicts, males almost always began with the least dangerous behaviors and only then stepped up to tail beating, and then, if necessary, to mouth wrestling, and occasionally on to circling. What's more, prior work with these cichlids had found that they are quite good at assessing weight asymmetries, and, as the model also predicts, the fights that Enquist and his colleagues recorded took longer when fish were evenly matched for weight.[15]

The sequential assessment model was off to a good start. But just how good were its predictions when it came to creatures other than cichlids? Enquist and Leimar invited Steve Austad to join them and find out.

Steve Austad was getting used to people contacting him out of the blue with unusual invitations. An animal behaviorist at the Department of Organismal and Evolutionary Biology at Harvard, he had done work using bowl and doily spiders (*Frontinella pyramitela*) to test another game theory model of aggression, called the war of attrition. While he thought his studies were of interest above and beyond what they said about power in these small, drab spiders, who take their name from their odd webs that look like an inverted cup and saucer, he hadn't expected to be invited to speak about that work at an international nuclear policy symposium. But that's precisely what happened.

That meeting was organized by Thomas Schelling, an economist who would go on to win a Nobel Prize for his work on conflict and cooperation. Schelling had told Austad that game theory models needed to be tested in simple systems as well complex ones, and that the spiders fit the bill for the former. So Austad went and stood before a group of experts on nuclear weapons, most of whom had no more experience with spiders than he had with atomic bombs. "I talked about the spiders," he says. "I spent at least a third of the talk apologizing for even being there. They listened and ate it up," he continued while laughing. "I kind of got worried about the hands that nuclear strategy was in."

Austad knew the sequential assessment model and Enquist and Leimar's tests of it in cichlids. But he was not quite sure what to make of it when one day in the early 1990s, Enquist called him from Stockholm and asked if he'd like to use the data he had collected on his spiders to test the predictions of the model. "To tell you the truth, I never took this too seriously [at the start]," Austad recalls. "They needed pretty much the highest computer power in the world to run their [mathematical] simulations . . . and I am thinking . . . there is no way this can be replicating any kind of intrinsic calculations in the spiders." Still, it was worth a try, especially since Enquist and Leimar offered to fly him over to spend a week reanalyzing his data to test parts of their model.

The data were based on 304 contests Austad had already staged between bowl and doily spiders in his lab. First, he had placed a female in a plastic container with some substrate for a web, which she quickly commenced to spinning. In 200 of the contests, he placed a single male in the container with her. After they mated, he added in a second male. In another 104 contests, he placed both males in the container simultaneously after the female built a web. In all cases, he sat and watched, taking notes on how contests played out. "They go through this grappling behavior," as Austad describes it. "They twist and tussle. When they grapple, they lock jaws and lock legs. And usually at some point, one of the males breaks away and runs away." Though the data did not allow for all the tests of the model that the work with cichlids had, they did allow Austad, Enquist, and Leimar to test the prediction that fights should last longest between equally matched opponents. "The model came close," Austad says. "I was surprised it replicated anything. Magnus and Ollie were trying to be quantitative and I thought this was asking too much of the spiders. I was absolutely surprised at how well [it worked]."[16]

Power is sometimes about control of place, and for animals that often means control of a territory. Any behaviors that reduce uncertainty about who controls access to a territory, and to the resources within it, will often be favored by natural selection. Such behaviors benefit the powerful, who waste less time assessing and reassessing the boundaries of their space, and the less powerful, who learn which areas are occupied and which are not.

Perri Eason has been thinking about power and what she calls "tactical defensibility" since her days in graduate school at the University of California, Davis, in the early 1990s. Back then, her dissertation work was looking at the social behavior of red-capped cardinals (*Paroaria gularis*) in Peru. "There were pairs [of birds] along the shores of a lake," she recalls. "I had a small sample size, but it looked to me like their boundaries were set up along landmarks, and I got interested in the nature of boundaries. I was interested in this for the next twenty years . . . but out in the field it is hard to find examples that would work cleanly." She got her chance in 1994, when one day the phone rang in her office at the University of Louisiana Monroe. The woman on the other end was hoping someone from the Department of Biology would come over and look at the wasps in her yard. Though wasps were not an area of expertise for her, Eason heard the voice of someone "really sweet, and I felt sorry for her and she seemed like she must be really lonely . . . to mollify her I said I would come look at her wasps."

When she got there, she saw five hundred cicada-killing wasps (*Sphecius speciosus*), "all defending territories in her yard. . . . It was crazy . . . and just for fun while I was standing there . . . I threw down a stick that had fallen off a tree to see what would happen and immediately two male wasps started defending either side of that stick." They seemed to crave boundaries to demarcate spheres of power. In no time, Eason had an experiment up and running in that very same manicured backyard. Her host was delighted: "I think she was just happy that someone was interested in *her* wasps," Eason says, "and that somebody else thought they were cool. . . . She brought me lemonade every day."

Eason began by laying out a grid on the yard, placing tiny stakes tipped with green markers into the ground every meter. Then she captured sixty-two wasps, chilled them, and marked their abdomens with unique color identifiers using enamel paint. Next, she mapped out the territories wasps already had established on the lawn by monitoring their chases and patrolling flights. A trip to the hardware store followed, where Eason bought thirty dowels and placed them haphazardly in the yard to see if the wasps would use them as territory markers. The wasps did not disappoint. All thirty dowels were used as boundary markers.

But Eason wanted more. She wanted to know *why* the wasps are so keen on using landmarks as boundaries. What do they get out of it? She hypothesized that landmarks reduced the costs of defending a territory, because the homes of the powerful were clear and obvious to all, and hence the time and energy spent on assessing what's what could be kept to a minimum. To test this idea, she returned to her new friend's lawn and put down fifteen pairs of dowels. Each pair was laid out with the two dowels parallel to each other, about 3 feet apart. Not surprisingly, the dowels were quickly adopted as boundaries, and soon there were fifteen new, well-defined territories between them. Eason had placed the dowels so that each of the wasp territories that she knew would be surrounded by four adjacent territories. For any given territory, two of the adjacent territories would border a dowel and two would not. Then Eason sat back and watched how the wasps spent time on territorial defense. As she had surmised, they spent significantly more time and energy on aggressive interactions with wasps on the adjacent territories that were not bordered by a dowel, and where an assessment of boundaries became necessary.[17]

A year later, Eason moved to the University of Louisville. Though she left the wasps and the lemonade behind her in Louisiana, her interest in power, boundaries, and territories came along to Louisville. For a variety of reasons, she set up her lab to study social behavior in fish, and in 2003—the same year that theoreticians Michael Mesterton-Gibbons and Eldridge Adams published a game theory model showing that, under certain conditions, natural selection would favor accepting landmarks as boundaries—she did some early laboratory experiments on boundaries and landmarks using blockhead cichlids (*Steatocranus casuarius*). But she yearned to test her ideas in the field again. Blockheads are native to Africa, but there are plenty of cichlids (somewhat) closer to Louisville, in the volcanic crater lakes of Nicaragua, so that is where she and her graduate student Piyumika Suriyampola headed.

They knew they wanted to look at landmarks and territories, but in *which* cichlid? "I just went to Nicaragua," Eason recalls. "There were quite a few [cichlid] species there and I thought something will probably work." She had been advised that one species, she doesn't recall which, would be perfect; however, it turned out to be anything but: "I collected two of them," Eason says, while laughing, "the whole time I was down there."

Their first sortie down to Lake Xiloá was in March 2011, and after a few days of scuba diving, Eason and Suriyampola settled on a cichlid species (*Amatitlania siquia*) that was perfect for what they had in mind. Their plan was to study how breeding pairs set up their natural territories before starting to experimentally tinker with those territories. For six days, each morning and afternoon, Eason and Suriyampola suited up in their scuba gear and began searching out breeding pairs. After allowing a pair to get used to them for five minutes, they began mapping out that pair's territory by observing where the fish would chase off an incoming intruder. They also took notes on their underwater writing pads about any obvious landmarks—rocks or beds of algae—at territorial borders. Then it was on to another breeding pair. What their survey found was that territories that had more landmarks tended to be smaller, and that, perhaps because territories without landmarks were less well-defined, in the eyes of both residents and intruders, pairs in areas with fewer landmarks chased intruders farther away from what they claimed as their rightful home than did fish on territories with more landmarks.

A year and a half later, Eason and Suriyampola were back at Lake Xiloá, this time with empty tin (beer) cans, plastic plants, and an experimental protocol in hand to look deeper into the role boundaries play in the power struggles of their cichlids. The empty cans, with their tops removed, served as a nest for breeding, and the plants, similar in shape to the native algae of the lake, as landmarks. Near some of the cans, they placed a single plant; near others, four plants in a line; and near still others, no plants at all. Within twenty-four hours, pairs began forming territories around the cans. When there was no artificial landmark present, they constructed circular territories, with the breeding can/nest at the center of each territory. If either type of landmark was present, the territories were smaller, and the fish used the plant(s) as a territorial boundary, which meant the breeding can/nest was close to a territorial boundary rather than at the center of a territory.[18]

Thinking back to both the cicada killers and the cichlids, Eason says it still amazes her "how responsive they are to landmarks. It was essentially the same thing. . . . You put down a stick for the wasps and they immediately use it. You put down a rock [or a plant] for a fish and they also immediately adopt it, like they were both hungry for the thing that wasn't there. . . . I think people would be amazed at the huge impact that such a small thing . . . has on what [an animal] is doing."

Landmarks or no landmarks, once territories are established, respective spheres of power are in place; neighbors know each other, having gone through a process of mutual assessment. A neighbor may still, on occasion, mount a challenge or takeover attempt, but at least it knows its opponent going in.

More than sixty years ago, British ornithologist James Fisher proposed that familiarity between established neighbors is a selective force in its own right, because neighbors are "bound firmly and socially, by what in human terms would be described as a dear enemy or rival friend situation," creating an uneasy alliance among those in power. Dear enemies are tolerated, in part because of prior assessments. Strangers, and the new assessments and power struggles they would create, are not.[19]

Early studies on dear enemies, like those of ecologist Robert Jaeger at the State University of New York at Albany, focused on how a territory holder responds when it encounters a particular neighbor (a dear enemy) versus a stranger. Jaeger studied red-backed salamanders (*Plethodon cinereus*), who navigate their world, including who is who in that world, primarily through odor. He combined field observations on territoriality with controlled experiments in his laboratory, measuring aggressive behavior by territory owners interacting with either a known neighbor or an unfamiliar red-back. The salamanders attacked and bit strangers more often than they did dear enemies. Bites were directed primarily at the snout, with its nasolabial grooves that are critical for hunting prey by smell; follow-up work detailed how such bites reduced the ability of victims to forage. The second most common bite site was the tail. These bites sometimes led to autotomy, in which the victim sheds its own tail, and with it, the substantial fat reserves within it.

Red-backed salamanders treat their *immediate* neighbors as dear enemies. But in some social systems, like that of the skylarks (*Alauda arvensis*) that Elodie Briefer studied, where many individuals in a locale share a common bond, who is considered a dear enemy may be more liberally construed, producing something akin to a dear neighborhood effect.

Briefer first began studying the vocalizations of skylarks while working on her master's thesis at Université Paris-Saclay, near Orsay, France, in 2005. "I wanted to study animal behavior in the field," she says. "I had done a small project on parrots, and I ended up talking with people at the bioacoustics lab. . . . I asked if they had any work for me, and they had a skylark project." Her master's work was an extension of work by Thierry Aubin, the head of the bioacoustics lab, who twenty years earlier had studied skylark vocalizations. "[Thierry] hadn't touched it again," Briefer adds, "and I think he just wanted to have someone working on it." That was a more than sufficient reason for an aspiring master's student to get a foothold in a potentially interesting system. Soon that work extended into a PhD dissertation project.

Each day during the skylark breeding season, from the end of February to late June–early July, Briefer headed out to study the skylark population near the university. She knew she had it relatively easy as far as fieldwork goes. Her subjects, tiny birds that are about 6 inches long and weigh under 2 ounces, were nesting in the beautiful fields near Orsay. What's more, Briefer says with a smile, "they sing [only] between nine and eleven in the morning, which was nice, and also they don't sing when it is raining or too windy. . . . I would drive along in my little car, and find places relatively quiet, not so close to a big road or airplanes . . . and when I found a good one, I would just go back every day to record them."

Her initial work focused on mapping out territories and deciphering the birds' vocalizations. Skylarks nest on the ground, and males fly over their territories, which meant that she could map out territories by watching where a male flew. Using an omnidirectional microphone with a parabolic reflector mounted above it, Briefer could also tape the song—composed of many different sounds or syllables—that a male was singing as he circled in the sky above his territory. Then she analyzed the songs on a spectrogram, measuring the durations of syllables, the durations of silences between successive syllables, the durations of sequences, and various attributes of "phrases," sequences of different syllables repeated over and over.

Briefer quickly figured out that skylarks were living in neighborhoods made up of small clusters of territories that were separated from other neighborhoods by a few kilometers. That sort of social structure is not unusual, but what was unusual was that the birds in each of the five neighborhoods she visited sang in their own distinct dialect: "shared sentences that are repeated in their song" is how Briefer describes these dialects. "They are typically 70 syllables long and shared by males of a given area. . . . If you go 2 kilometers away, there are no shared sentences. You have the same units [syllables], but not arranged in the same sequence."

What was also very clear from Briefer's observations and her spectrogram analyses was that the birds "react strongly to strangers . . . not so strongly to neighbors." And what defined a stranger to a breeding skylark seemed to be not whether its territory bordered his own, but whether it sang the dialect of his neighborhood. Perhaps, Briefer reasoned, a dear enemy effect at the level of the neighborhood was in play: that in addition to a personal sphere of power in the territory, assessment of song in neighborhood birds created a *shared* sphere of power.

To test that idea, Briefer and her colleagues ran a series of playback experiments in which they broadcast ninety-second snippets of songs of different birds to a territory holder. Three different kinds of songs were played to a territorial male: a song from his neighborhood, a song from a different neighborhood, or a chimeric song, made up of a song from a different neighborhood into which Briefer had inserted parts of the territory holder's dialect. These songs were broadcast from a speaker placed 5 meters inside the territory of a male as he flew over that territory, while Briefer observed his behavior toward any other male flying in the vicinity of his territory, as well as how he behaved when he landed and was near the speaker.

What she found confirmed her initial sense that skylarks use dialect to categorize others as neighbors or strangers. A male was more likely to chase away another individual flying near his territory when he heard the song of a bird from a different neighborhood (or the chimeric song) than when the song was from a neighbor. When on the ground, a male was also more likely to approach a speaker if it broadcast the song of a skylark from another neighborhood, which is consistent with the kind of assessment that occurs before chases. Briefer found that it didn't matter if a song was that of a neighbor from an immediately adjacent territory or from farther away in the neighborhood—everyone in the neighborhood was treated as a dear enemy and not a stranger.[20]

Power is not *always* communally distributed among skylark neighbors, however. The neighborhood-level dear enemy effects that Briefer and her team observed are tempered by changes in the ecological and behavioral environment that take place as the skylark breeding season progresses. The team found that from late February to mid-April, when males were establishing territories, no neighborhood-level dear enemy effects were in play: males treated everyone as if they were strangers as territory formation unfolded. By mid-May, when territories were established and a first clutch of eggs was laid, the neighborhood-level dear enemy effect was in full force. Others from the neighborhood were tolerated; strangers from other neighborhoods were not. Then, after the second clutch of chicks hatched in late June, the birds went back to treating everyone as a stranger. That last finding initially puzzled Briefer, but eventually she reasoned that young birds leave the nest, first walking about and then learning to fly, starting in late June. All this traffic leads to confusion, and the constant movement of active chicks across territorial boundaries, plus parents chasing after them, may make territory holders more sensitive to *any* skylark coming near their sphere of power.

The strategic assessment of rivals for power seen in caribou, hermit crabs, cichlids, spiders, wasps, and skylarks is ubiquitous in social animals. Information about potential rivals is paramount in the struggle to secure and then maintain power. As we're about to learn, every bit of information an animal gathers is critical intel: whether you've recently lost or won a fight, whether your current opponent has won or lost recently, even whether you are being watched, and more. Everything matters when it comes to power.

4   Watch and Be Watched

> The more I come to know people, the better I like ravens. If I have
> a religious practice, it is the watching of these birds.
> LOUISE ERDRICH, *The Painted Drum*

Anyone who has watched ravens (*Corvus corax*) being ravens needs
no convincing about how sociable and intelligent they are. Thomas
Bugnyar, an animal behaviorist now at the University of Vienna,
had certainly seen his share of these remarkable birds, but he never
expected he'd spend the better part of his life studying just how
sociable and intelligent they are, nor how those characteristics play
into raven power structure. Then he went up to the Northern Austrian Alps one day. "A friend of mine was part of a team hand-raising
ravens at a field station [there]," Bugnyar recalls. "I was quite impressed by the birds, because they were not actually behaving like
birds, but more like dogs and puppies."

If that first visit wasn't enough to grab his attention, not long
after that, when he had joined the raven research team himself, one
of the birds made him a true believer in all things raven. The ravens
he was working with were housed in pairs, and a raven in one pair
had escaped. The other one "was all by himself for a few months,"
Bugnyar says, "and during that time I spent every lunch break with
him, playing with him, just to give him a little contact, and I also en-
couraged everyone else to interact with him, because he was bored."

Bugnyar would occasionally give his raven friend a piece of cheese as a snack, and one day he had a small slice in his pocket and reached in to get it as the bird sat on his arm. "I showed him the piece of cheese," he continues, "and as soon as he saw it he went for it in a very fast way, and so I pulled back my hand, because he came so fast. The beak is big and if it does not hit the cheese, but your finger, it hurts." The raven looked him squarely in the eyes and said "aua!" which happens to be the German equivalent of "ouch!"—an exclamation the raven had no doubt heard used by one of the many humans who had befriended him. "I said, wait a second, 'no, almost aua,' you behaved yourself," says Bugnyar. "My interpretation was that he anticipated my expression ['aua'] in response to his behavior. . . . He was using [a human] expression in the perfectly right context. . . . His social environment at that time was humans and so he was trying to make sense of what we were doing. It also illustrates quite nicely why I ended up using vocalization as a window into what they actually think." And, he would come to learn, into their power structure as well.

When he began his PhD work with ravens in the mid-1990s, Bugnyar could hardly have found a more picturesque place to work. The Konrad Lorenz Field Station, about 230 kilometers east of Vienna, sits near the bucolic village of Grünau im Almtal in the Austrian Alps. The station, which is home to hundreds of ravens, many of whom have been banded and individually tagged, sits in a valley and is part of a game park. The park is a wilderness full of wolves, bears, deer, and more, where, Bugnyar says, "from time to time you see an enclosure."

Some of the park enclosures also house wild boar, and they're fed daily by the park staff, always with plenty of leftovers for the ravens to gobble up their (uninvited) share as well, which means Bugnyar knows where birds will be and when they will be there. He's habituated the ravens to the presence of people and can get within 10 meters of them, and because many of them have been marked, Bugnyar's team also know who's who. Some of the marked birds at the park have been there going on fifteen years; others live a more vagabond life.

Year in and year out since graduate school days, Bugnyar has re-turned to the station to continue his now 25-year-long study. Armed with binoculars and tape recorders, he and his students have seen—and, just as importantly, heard—thousands of power-related inter-actions. Struggles for power take many forms in ravens, including approach-retreat sequences, in which one bird backs off as soon as another approaches; forced retreats, during which a raven retreats after being threatened; and true fights, when birds grapple using their very sharp beak and claws.

From the birds' perspective, Bugnyar and his team are an audi-ence that is not worth paying any mind to. But audiences made up of other ravens are a different matter. Ravens who are victims of ag-gression are known to give "defensive calls," and prior work by Bug-nyar and his team found that bystanders sometimes come to the aid of victims who utter these calls. Bugnyar concluded that "the caller wanted help and to [convey] the idea 'I have a problem now.'" But it seemed to him that there was more going on. "Sometimes when there is a beat-up, [the victim] seems to cry like mad . . . even if it is a mild beat-up . . . they seemed to me to overdo it," he says. "At other times, there is a pretty intense beat-up and they seem to stay quiet." He began thinking that the composition of the audience of ravens who were watching and listening might be what made the difference. In 2010, funded by a large grant titled "Raven Politics," Bugnyar and his graduate students Georgine Szipl and Eva Ringler decided to dig deeper.[1]

The team videotaped forced retreat interactions in which the vic-tim uttered a defensive call and noted the duration and number of those calls. They also collected information on the identities of other ravens within 25 meters of the interaction. Using their long-term database, they then classified each bystander as kin (or not) to the victim or aggressor. They also scoured their records to see whether a bystander had a strong social bond with the victim or the aggressor, as measured by whether the two were mating partners or had ex-changed affiliative behaviors such as grooming.

Bugnyar and his colleagues found that ravens on the wrong side of a power struggle modulated their defensive calls depending on the nature of the audience. Victims' call rates were higher when potential allies—either kin or those with whom the victim had a strong bond—were audience members. But there were other considerations, too. The victims were not only attuned to who might have a proclivity to help them, but were also taking into account whether the audience was made up of those that might make their situation even worse by coming to the aid of the aggressor. When the audience was made up of individuals who had a history of prosocial behaviors with the aggressor, victims, who might suffer by drawing even more attention to their unfortunate predicament, reduced their call rates.[2]

Ethologists have come to learn that paying attention to who is in the audience is just one of the many tools that animals use to help them acquire power. We will discover that gathering intel in all its forms should be, and is, favored by natural selection. Being watched is one thing; spying on others in their quest for power, as the swordtails we discussed in chapter 1 did, is another. Little blue penguins in New Zealand will remind us just how much intel that sort of behavior can bring in. But it isn't just what others are doing or have done that provides important information: monitoring oneself, even if not at a conscious level, does as well, as we will discover in our sortie into winning streaks in copperhead snakes in Michigan and losing streaks in California mice and white-footed mice.

But first, a bit more on audience effects and power: this time in chimpanzees.

From the time he started his master's work studying the alarm calls of Diana monkeys (*Cercopithecus diana*) in the forests of Ivory Coast, Klaus Zuberbühler has been fascinated by power and vocalizations in primates. When he landed a job at the University of St. Andrews in Scotland in 2001, Zuberbühler assumed he would continue that work, but a coup in Ivory Coast led to his students being airlifted to safety and brought an abrupt end to the project. Fortunately, he was also working with some people at the Edinburgh Zoo, and they had connected him with primatologist Vernon Reynolds, who ran a long-standing study of chimpanzees (*Pan troglodytes*) in the Budongo Forest in Uganda. Reynolds was about to retire and essentially handed his system off to Zuberbühler.[3]

Zuberbühler and his team have been studying two chimpanzee groups at Budongo. Each group is followed every day, from 7:00 a.m. to 5:00 p.m., by three or four Ugandan field assistants, who record behavioral data on a handheld computer. When Zuberbühler and his team are doing experiments, the logistics of working with agile, smart creatures who are going to do what they want, not what you want, can lead to mind-numbingly long waits: "People sometimes follow their focal animal for a week to finally get one trial done," Zuberbühler says with a smile, ". . . but the reward is great because if there is a pattern that comes out with experimental [field] data, it just tends to have a lot of power scientifically."

One of the things that became obvious very quickly is that chimpanzee power struggles are loud. Both aggressors and victims scream, and what struck Zuberbühler was that "depending on who is listening, they tend to exaggerate, more or less, about the nature of the attack." Just like the ravens. He began to think that the nature of the audience was the key to these vocalizations. "If you are being attacked . . . [often] the only way to get out of it is to get someone else to join and that may turn the tide. . . . If the [victim's] scream recruits help, then it really matters who is nearby. Especially if it is the alpha male, who does not tolerate violence among others."

When Zuberbühler and Kate Slocombe analyzed eighty-four chimp power struggles, they found that when fights involved only mildly aggressive interactions, the animals took no account of whether an audience was present. When contests involved severe aggression, victims' screams were longer and more intense when an audience was nearby—but only when at least one of the audience members held a rank in the dominance hierarchy that was equal to or above the rank of the aggressor. This strategy seems to pay: victims that emitted longer and more intense screams received support from high-ranking observers who often intervened and broke up fights.[4]

Chimps and ravens are hardly alone when it comes to the role an audience plays in power dynamics. Audience effects have been found in Japanese quail (*Coturnix japonica*), fiddler crabs (*Uca maracoani*), zebrafish (*Danio rerio*), and Siamese fighting fish (*Betta splendens*). In fighting fish, it gets particularly interesting. A male's testosterone level changes when he is being watched, and some work suggests that fighting males change their behavioral repertoire if a female, but not another male, is watching, and that all of these changes are mediated by whether a male knows those in the audience.[5]

Audience effects are just one of a class of power-related phenomena called extrinsic effects. Extrinsic effects, in contrast to intrinsic effects such as size and weight, incorporate various aspects of the experience and the social environment of those seeking power. In addition to audience and bystander effects, two other types of extrinsic effects are winner effects and loser effects. Winner effects occur when an animal's chance of winning a contest for power increases as a function of prior wins. When the probability of losing increases as a function of prior losses, loser effects are in play. Of the two, loser effects seem to be more common, and ethologists like Gordon Schuett have turned to copperhead snakes to better understand just why that is.[6]

Most teenagers want to turn their basements into game rooms, not laboratories housing dozens of venomous snakes that they've caught. But most teenagers are not budding herpetologists like Gordon Schuett was back in the day. "I was very fascinated with venomous snakes as a young boy," Schuett says. "I was collecting rattlesnakes when I was fifteen." Soon he was not only milking the rattlesnakes for their venom, but as a high school student was reading the primary scientific literature on all things venomous snake. The mini-lab in the basement was in no small part due to his very busy, but equally open-minded, mother.

At sixteen, Schuett became obsessed with power in copperhead snakes (*Agkistrodon contortrix*). "I was utterly struck by male-male combat [and thought,] 'I'm going to try and do that in my basement.' . . . And lo and behold, I was able to get them into combat." And four decades later, he's still enthralled by power struggles in venomous snakes.

After doing some undergraduate work on sperm storage in female copperheads at the University of Toledo, Schuett caught the attention of Jim Gillingham, one of the leading experts on copperhead behavior at Central Michigan State University, just a stone's throw from where Schuett grew up. He gave Gillingham a tour of the lab he still had in his family's basement, and before long he was starting a master's project quantifying power contests in his beloved copperheads. To get his subjects, Schuett would scour the roadside after summer evening rainstorms. "You have a snake hook or a grabber and the snakes go in a bucket," he says, then adds, as only a herpetologist would, "Copperheads are relatively inoffensive . . . only about a meter long."

Copperheads are solitary creatures, except in the late summer and then again in the spring, when they gather in large mating aggregations. And that's when fights are aplenty, so Schuett's timing worked well, as the copperheads he had brought in from the field were acclimated to the lab at Central Michigan State by the time he started observing contests in the late summer.

When they fight, copperheads live up to their species name, which translates from Latin roughly as "twisted fishhook." During contests, males contort themselves in every conceivable way, angling for leverage. Power contests typically begin with challenge displays, including "ascend," in which a snake rises up from the ground, and "sway," in which it undulates back and forth. Sometimes these displays result in a snake flashing "head hiding" behavior before retreating, with the winner following in hot pursuit. If the challenge displays don't settle the contest, a male sometimes tries to "hook" his opponent, positioning himself above and around it. If he is successful, the opponent is often forced to the ground. If snakes are simultaneously hooking each other, they intertwine and stiffen, eventually separating, with one male emerging as the victor and the other showing submissive behavior and retreating.[7]

Initially, Schuett watched these power struggles from behind a blind in the lab, but soon realized that was unnecessary: "They didn't care if I was there," he says, laughing. "When they wanted to fight or court, they did." As he watched, it struck him that losers not only retreated, but went into what he called a refractory period, during which they avoided aggressive interactions with anyone. Schuett had read about such things in other species — mostly fish — when he was doing this work in the early 1980s, but in those studies the effect lasted hours, while in his snakes the refractory period seemed to last a week or more.

With a detailed inventory of copperhead aggressive behaviors in hand, Schuett moved on to start his PhD with David Duvall at the University of Wyoming. He wanted to go one step beyond his prior work on losing and measure its effects on reproductive success, so he ran a series of experiments in which two copperhead males were paired near a female during the late summer mating season. In the first of these experiments, neither male had had a recent losing experience, but one male was about 10% larger than the other. Thirty-two such triads were formed, and the larger male won every single time, subsequently courting and guarding the female. Twenty-four hours after the initial battles were over, Schuett took ten of the winners and ten of the losers and pitted each one against a similar-sized male who had no prior fighting experience. What he discovered was that copperheads don't go on winning streaks: first-round winners were no more likely to win than were their opponents. But losing is very bad news. Males who lost in the first round were *never* the first to challenge their next opponent, and in each and every fight, they retreated. The winner then courted and guarded the female. When Schuett ran the same sort of experiment seven days after a loser had lost, he found very similar results. Losing begat more losing.

Schuett wondered what would happen if he paired a snake that had recently lost against an opponent that was 10% smaller. Would the size advantage, which was clearly part of the story but not the whole of it, compensate for the prior loss? The answer was a resounding no: the losers lost again. Lose once in the battle for copperhead power, and the effects linger and create a serious fitness cost. But why would it be adaptive to effectively shut down the quest for power for a week when the whole mating season lasts for only a month? One possibility is that the life span of a copperhead is long enough that things might get better at some point down the road, so it pays to temporarily shut down. "If you engage in combat and become the loser," says Schuett, "in theory you have lost one-quarter of the mating season. If you lose again, you may never want to engage in courtship again that season. I think they can only manage that burden because these animals can live twenty-five or thirty years."

Still finishing up his PhD, Schuett was pleasantly surprised when a talk he gave on loser effects in his copperheads was picked up by the *New York Times*. As if that wasn't enough to excite a young scientist, "two weeks later Carl Sagan writes me," he recalls fondly. "The letter is still on my desk and [the study] ended up in his book *Forgotten Ancestors*."[8]

Excited by all that the copperheads were teaching him about power, Schuett next turned to ask what was causing loser effects in real time—that is, at the physiological level, what was making losers more likely to lose again? To find out, he and his colleagues again put pairs of males together near a female and waited until one male was the clear winner. At that point, they separated the males and took a blood sample from each. For comparison, they also ran two controls: in the first, they took a blood sample from a lone male; and in the second, they placed a single male and a female in the arena, then took a blood sample from the male. Analysis of the samples revealed that plasma corticosterone, a key stress hormone, was significantly higher in losers than in winners or control males, suggesting that increases in stress hormones might signal to loser males that they should shut down the quest for power and wait for better times.[9]

Cathy Marler knew of the work on loser effects in copperheads. Indeed, she could find many examples of loser effects in the animal behavior literature. But what of winner effects? Occasionally she came across such effects in a study, but any boost from winning seemed short-lived. While it was true that some mathematical models suggested loser effects were more likely to evolve than winner effects, Marler felt something was missing. It just didn't make sense to her that winner effects should be transient when they seemed so important in helping animals gauge the social environment and determine when to ramp up aggression.[10]

By the time she had familiarized herself with the literature on winner and loser effects, Marler had already done extensive field-work on aggression in both lizards and frogs, and had developed a special interest in the hormonal and neurobiological underpinnings of power. When she took her first tenure-track position in the Department of Psychology at the University of Wisconsin in the late 1990s, she was looking for a species she could work with in the lab. On that count, she got lucky. Just as she arrived at the University of Wisconsin, another faculty member, who worked with five species of mice, was leaving and looking for someone to take over his soon-to-be-orphaned rodent colonies. Marler jumped at the chance. She had hopes of working with all five species, but when that turned out to be too expensive, she decided to focus on just two: the California mouse (*Peromyscus californicus*) and the white-footed mouse (*Peromyscus leucopus*).

The choice of California mice and white-footed mice was anything but random. "I love diversity of behavior," Marler says, and for someone with such proclivities, it is hard to imagine two species that are so closely related, but have such different social systems. Both look like your standard mouse of the field, but California mice are monogamous, while white-footed mice are polygamous, and male and female California mice provide more parental care than their white-footed counterparts. California males are also more aggressive to intruders than are white-footed males, which may in part be due to their having more brain receptor sites for a hormone called arginine-vasopressin (AVP), known to promote aggression in male, but not female, mammals.[11]

To dig deeper into AVP's role in shaping aggression and the path to power in these two species, one of Marler's first experiments was a cross-fostering study. Cross-fostering experiments with two species involve raising the offspring of individuals of species 1 in the nest of species 2, and vice versa. If offspring behave like their foster parents when they mature, this result suggests that the developmental milieu strongly affects behavior.

Together with Janet Bester-Meredith, Marler raised twenty-four white-footed pups with California foster parents and fourteen California pups with white-footed foster parents. When foster offspring were about seven months old, they were tested in standard aggres-

sion tests that had been developed for rodents. They generally behaved more like their foster parents than their biological parents.
The strongest effects were found in California males, who showed
much lower levels of aggression when they grew up in the nest of
white-footed foster parents. This change in aggression was partly
due to the effect of cross-fostering on AVP brain receptors, as foster
California males had both fewer and smaller cells at AVP receptors.
These results suggest that the road to power can be rerouted when
normal developmental patterns are altered.[12]

In the early 2000s, with an understanding of some of the basics
of power in both *Peromyscus* species, Marler, working with a team
of students, set her sights on examining winner effects. In these experiments, a male and female were housed together in a cage, and
the male was given zero, one, two, or three winning experiences. The
winning experiences were provided by allowing the resident male to
interact with a smaller, weaker, semi-sedated male placed in his cage.
Next, a healthy intruder, about the same size as the resident male,
was placed in the resident's cage. Marler and her team observed the
fights, and when the contests were over, they drew blood samples.

In white-footed mice, although winners had lower stress hormone levels than losers, no winner effects were uncovered; even
when a male had just come off three wins in a row, it did not affect his chances of winning against the healthy intruder. And winning had no measurable effect on testosterone level, either. In California mice, which rely more heavily on aggression in their quest
for power than do white-footed mice, winning matters—but only
when it's winning on a grand scale. Males with one or two prior wins
were no more likely to defeat an intruder than were males with no
winning experience. But if a California mouse had really been on a
roll and had had three wins, he was likely to beat any intruder who
entered his domain. This "third time's the charm" effect was linked
to increased levels of testosterone in males coming off a string of
wins. Together with two of her students at the time, Matt Fuxjager
and Elizabeth Becker, Marler was even able to pinpoint the brain
circuitry associated with changes in testosterone and winning. But
winner effects and the rise in testosterone depend, as many power-
related things do, on place and ownership. When Fuxjager and Mar-

ler ran an experiment similar to the previous one, except that males had to fight outside their territory, the winner effect and associated rise in testosterone were much less pronounced.[13]

Fuxjager and Marler thought more about the differences in winner effects between California mice and white-footed mice, and the role that testosterone plays in mediating these effects. Was it the case that white-footed mice, who showed no winner effects, lacked the physiological machinery to mount a winner effect, or could it be that they have that machinery, but just don't produce enough testosterone to kick it into gear? They began to wonder what might happen if they experimentally increased testosterone levels in white-footed mice to the levels typically seen in California mice displaying winner effects. Could they mimic the winner effects found in California mice?

Fuxjager and Marler tested thirty-seven white-footed mice in three groups. In one group, males experienced three wins (by being matched against much smaller opponents), and after each win they received an injection of testosterone. Two control groups were also tested. In the first, males that had experienced three wins in a row were injected with saline, to make certain that it was an injection of testosterone, not just any injection, that was causing a winner effect. In the second group, males experienced three wins, but no injections. The testosterone injections appeared to be a power elixir, as white-footed mice that won three times and received testosterone each time showed the same winner effect that seemed, to that point, reserved for California mice.[14]

Schuett's and Marler's work has shed new light on loser and winner effects, but there is another extrinsic effect that we touched on in our earlier discussion of swordtail spies that leads us now straight to ethologist Joseph Waas lying flat in penguin poop night after night in a New Zealand cave.

In 1983 Waas, a birder to his core, was finishing up his undergraduate work at Trent University, in his native Canada, and was on the hunt for what to do next. "At that time in New Zealand," Waas says, "you could pretty much pick a bird and be the first person to study its behavior." John Warham, a pioneer in the study of penguin biology, suggested he go check out a colony of little blue penguins (*Eudyptula minor*) on the eastern side of Banks Peninsula. "I went . . . and they just fascinated me," Waas recalls. "It seemed so cool these penguins lived in caves and in burrow communities and were active at night. And so I started working on their vocal repertoire."

The smallest of all penguins, standing only about a foot tall, little blues are as cute as can be. But they're loud. Really loud. "There would be these periods of time," Waas says, "when all the calling would die down and then one bird or maybe two birds would then call and then you would get this incredible contagious effect until everyone started calling." His early work was in the cave colonies, where penguins nest about 2–3 meters apart, usually up against the wall of the cave. Because they are nocturnal, Waas needed to be as well, arriving at the colony at dusk, in time to watch the penguins return from the ocean and waddle to their caves. He'd follow hot on their trail, with a tape recorder and a camera hooked up to a night-vision scope in hand, and stay in the cave with the penguins until 4:00 a.m. Then back to Christchurch for a bit of sleep, to start the whole round trip over again the next day.

It was not easy work. "The main cave I worked at is in Ōtanerito Bay," Waas notes. "There are two areas, an upper [cave] area that had maybe a hundred birds . . . and then there was a lower cave, where you had to get on your belly and crawl, which wasn't very pleasant as the actual base of the cave is made of dried guano and penguin feathers, all sort of smashed [up]. Dreadful smelling and not very pleasant to breathe." The face mask he wore helped a bit, but only just a bit.

In addition to vocalizations, Waas was also interested in power dynamics in these tiny penguins. Lying on his stomach in the lower cave, what he saw and heard fascinated and frustrated him: birds fighting, interlocking bills, flipping one another, as Waas says, "almost like a judo throw." His mind was soon racing with ideas, but how was he going to design an experiment? "What was I going to do?" he says. "I couldn't set up fights between penguins, so I put it on the back burner."

One thing Waas could not help hearing over and over on all those guano-soaked nights was the "triumph call" made by males. First described in greylag geese (*Anser anser*) by Nobel laureate Konrad Lorenz, the triumph call in little blue penguins is a high-pitched inhalation paired with a bray-sounding exhalation that is repeated over and over. But it was the context of the call, not its auditory dynamics, that really struck Waas: at the end of an aggressive interaction, the winner often stood up with its flippers out and "belted out this [triumph display] vocalization," while "the loser [would go] into a low walk or a low run . . . directly away from the winner." And sometimes it was doubly spectacular: if a female was on the nest of the male who was making a triumph call, she sometimes joined in.[15]

Waas knew he was not the only one impressed by triumph calls—other penguins, not only the male who had just lost to the caller, were clearly paying attention to these shouts of power as well. Exactly why, and what these eavesdropping penguins might be doing with the information from the call, though, he didn't know. Waas still could not—and even if he could, would not—set up fights between little blue penguins, but began to think he might be able to use playback experiments, where he could control what was heard, to study power and possible eavesdropping effects. But the cave he was working in was just too chaotic to control who would hear what, and so he turned to another colony near the cave, where penguins lived outside in burrows.

This burrowing colony was in the middle of a farm owned by two of Waas's friends, Francis and Shireen Helps. "They are alternative farmers," Waas explains, "and they made a real effort to maintain an area of their farm to protect the colony." Equally important in the eyes of an ethologist, "they marked all the birds, and so for a lot of birds we knew their age and sex, and they also constructed artificial burrows . . . that the birds seemed to prefer." Those artificial burrows were all exactly the same dimensions—300 × 350 millimeters in area and 200 millimeters high. So, all of a sudden, instead of the chaos of the cave, where any playback experiment on power structure and eavesdroppers was impossible, here was a marked colony, with birds living in identical homes—a perfect setup for Waas to run a playback experiment, putting out some speakers that allowed him to control who heard what. Working with Solveig Mouterde, a veterinary graduate student at the University of Waikato, he did just that.

In a paper in which he listed the Helps as coauthors—not only because of their foresight in protecting the birds, but because they welcomed Mouterde to live on their farm during the study—Waas and his team detail the experiment. They worked with twenty-seven males and sixteen females who were incubating eggs alone at their artificial burrows (both males and females incubate eggs) while their partners were off foraging in the ocean. The egg was gently removed and put into an incubator, and an artificial egg was placed in its stead. The artificial egg had sensors that recorded the bird's pulse and so, indirectly, its heart rate. Next, they put a microphone near the nest they were working with that night, so that they could record the vocalizations of the penguin after he or she heard a speaker 5 meters away broadcast the sounds of a fight, followed by the triumph call made by the winner of the fight or, alternatively, the calls made by the loser of the fight.

When males heard the triumph calls of winners, their heart rate shot up by more than thirty beats per second compared with typical baseline values. No such burst was found when they heard the calls of losers. These eavesdroppers were clearly nervous when those in power were nearby. And they acted like it, too, as they were much more likely to vocalize themselves in response to hearing the sounds of a loser, who would presumably be a weaker potential opponent, than the call of a winner. For their part, females showed increased heart rates when they heard winners or losers, and never vocalized after hearing playbacks, suggesting they were generally agitated by fighting and wanted no part of any of it.[16]

Eavesdropping and audience effects—and, to a lesser extent, winner and loser effects—tell us that experience with others matters in the struggle for power. But these effects always cast the other as an obstacle on the road to power. Alliances and coalitions, formed in part to acquire power, set others in a very different light.

5 Build Alliances

For how can tyrants safely govern home
Unless abroad they purchase great alliance?
WILLIAM SHAKESPEARE, *Henry VI: Part 3*, 3.3.69–70

The path to power can be jagged and dangerous. It helps to have allies by your side, to build and employ coalitions and alliances toward your end. Still, recruiting them is no easy task and requires skills in the realm of social intelligence.

When ethologists speak of social intelligence, they generally mean the ability to navigate in a socially complex society full of other socially complex beings, all of whom are trying to maximize their fitness as you try to maximize yours. This ability is distinct, some have argued, from a more general type of intelligence that involves solving problems like how to find food, build a shelter, escape from a predator, and so on. Much of the work on social intelligence has been done with primates, and while, of course, the specifics vary from species to species, primates not only have relatively large brains and distinguish kin from non-kin, but are also remarkably astute at using past interactions to predict future outcomes and adjusting their behavior accordingly. They have and use information about the power of others and how others wield that power, and individuals preferentially interact with partners that stand to provide them with the most benefits.[1]

The thing is that when Kay Holekamp went down this rather impressive list and asked whether hyenas—who, as we have seen, live in complex social milieus—show the same traits, she could check every box. And one way hyenas employ their social intelligence is to form coalitions and alliances to attain and maintain power.[2]

When Holekamp was a grad student, she had read an edited book on coalitions and recalled it later when she realized, "Wow, we see that in hyenas all the time." It seemed to her that coalitions served to reinforce the status quo, and that in her hyenas, "groupies were constantly trying to ingratiate themselves with the higher-ranking individuals . . . [and so were] in a lot of coalitions."

Holekamp and her team decided to take a deeper look at power dynamics in hyena coalitions. Working from their Fisi basecamp, they found that about 14% of the almost 12,000 aggressive interactions between hyenas they observed involved a coalition acting against a third party. Typically, an individual was involved in an aggressive interaction, and once that interaction had commenced, it was joined and aided by its coalition partner. Pairs of adult females joining forces was the sort of alliance most often observed. But toward what end? What benefits did hyenas derive from being part of a coalition? Holekamp first looked to the obvious, but there was no evidence that coalition members got better access to food. Instead, what she found was that because females in coalitions directed most of their aggressive behaviors toward those below them in the dominance hierarchy, they were, as she had first speculated, reinforcing the status quo. That said, coalitions were occasionally responsible for revolutionary change. The power structure in hyenas is usually quite stable, and rank is generally passed down across generations from mother to daughter. Sometimes, though, parts of the power structure are upended, and individuals unexpectedly, at least to Holekamp, move up in rank. It turns out that those revolutions are often associated with females who have joined in a coalition to attack and defeat those above them.

Another benefit of being part of a coalition was that other members tended to be genetic kin, so coalition members were increasing what animal behaviorists call their inclusive fitness. Classic fitness is measured in the number of offspring produced. But because genetic kin are, by definition, likely to carry the same gene variants, inclusive fitness not only counts the number of offspring an individual produces, but also credits an individual for the number of additional offspring its genetic relatives produce as a result of its help, including its help as a coalition partner. Actually, it's a bit more complicated than that, but the take-home point is that hyenas who form coalitions with genetic kin get a sort of indirect genetic kickback for doing so.[3]

Hyenas are hardly alone. Ravens are one of the quintessential masters of social intelligence. They, like hyenas, put that talent to use forming coalitions, but there's more to it than that. Thomas Bugnyar noticed that "when two birds are preening each other or playing, [sometimes] a third one comes [along] and says no, stop it." In one data set that he and his colleagues gathered at the Konrad Lorenz Field Station, about 18% of the 564 pairs of ravens they observed interacting in an affiliative manner suffered such interference from outsiders. Most cases involved the third party acting in an overtly aggressive way, but about a quarter of the time, it simply placed itself between the pair. Though it was not without risk—sometimes an interloper was counterattacked—about half the time this meddling worked, in the sense of breaking up the interaction between the friendly pair.

Bugnyar knew that ravens keep track of the dominance rank of others in the group. When he talked over these observations with his team, "[we thought] if your own friend was interacting in nice interactions with someone else, you don't like this, so you go and intervene." But when they dug into the data, they found no such pattern. Instead, what was clear was that the intervening bird was always a powerful, high-ranking individual. We'll look more at intervention behavior in the next chapter, but, with respect to coalitions, what matters is that high-ranking birds didn't always step in when others were displaying prosocial behavior. They were much more selective than that. "They completely ignored those that were already closely bonded and they completely ignored those that were not bonded,"

Bugnyar says, instead "selectively targeting those that were [in the process of] bonding." Dominant ravens appear to interpret nascent coalitions as a threat to their own power—and with good reason, because once individuals form a coalition, each bird tends to rise in the power structure.[4]

In so many species, in so many places, we see coalitions play a role in animal power structures. We see coalitions on the land and in the sea and in the air, coalitions in the bays of Australia, the forests of the Democratic Republic of the Congo, the meadows of France, the forests of Tanzania, a zoo in the Netherlands, and more. Researchers are trying to understand how and why coalitions form, and how they both create and alter power structures, using observational and experimental studies and a bit of mathematical theory to guide them. They ask, among other things, about the importance of kinship and reciprocity, and why females form coalitions in some species while males do in others.

Somewhere along the way, ethologist Richard Connor learned that if you take body size into account, the dolphin brain is second only to the human brain in size. "What were they doing with those big brains?" That question, as Connor will tell you, became his obsession. "I wanted to find a place," he says, "where you could actually watch wild dolphins and see what was going on."

His timing was good. One day when Connor was an undergraduate student at the University of California, Santa Cruz, in the late 1970s–early 1980s, Elizabeth Gawain, a local city planner, came over to the Department of Zoology, color slides in hand, to regale its members with a recent trip she had taken to Shark Bay, Australia. She talked glowingly of the incredible bottlenose dolphins (*Tursiops aduncus*) swimming there, just waiting for someone to come study them. Connor had heard a little bit about those dolphins from Rachel Smolker, another student in the department. He was one of the few undergraduates in the audience that day, Connor recalls, "and the graduate students weren't going to drop [their projects] to go out and run off to this possibility in Australia." But he was more than happy to drop everything to do just that, with the sort of expectations excitable undergraduates are prone to: "I thought it would at the very least be like the situation at Gombe . . . where Jane Goodall [was] watching social interactions at close range."

When Connor graduated in 1982, he sold his coin collection, got a small grant from the Explorers Club, and went off with Smolker to Shark Bay, near Monkey Mia, on the far western coast of Australia, about 850 kilometers northeast of Perth. They didn't have a boat and, indeed, Connor says with a smile on his face, "we didn't have much of anything." Still, on that first pilot sortie to Shark Bay, they managed to borrow a small craft. What they found were hundreds of bottlenose dolphins that, to their delight, Connor recalls, "didn't mind us being there, and we thought 'Wow!'" That wow translated into Connor entering the PhD program at the University of Michigan under Richard Alexander and Richard Wrangham, and Smolker doing the same. They found themselves smack in the middle of a hotbed of cutting-edge ideas on behavior and evolution, led by some of the founders of the nascent field of sociobiology.

In some years Connor was at Shark Bay, in other years Smolker was, but most often, both were there. They pitched a tent at the Monkey Mia campground. At the start, there was no road from the campground to Red Cliff Bay or the other spots they launched their boat from, though eventually a "road" of bitumen was laid, making things a bit easier. The amount of time they had with the dolphins was determined by how choppy the water was on a given day.

Shark Bay is large (90 by 50 miles), and along with the dolphins, it's teeming with snakes, turtles, sharks, dugongs, and the like. The average depth of the bay is 27 feet, but where Connor and Smolker did most of their observations, it tended to be about 48 feet, which is remarkably shallow water for dolphins, allowing for observations that would be difficult to make elsewhere. Initially they used a small dinghy that Smolker had purchased with National Geographic Society money, but soon they had moved up to a 16-foot craft. They usually sailed parallel to the Peron Peninsula in the bay, and never more than a few miles from the shore. With survey sheets, camera, and tape recorder in hand, and always standing up looking down starboard or port, they gathered data on the behavior of hundreds of dolphins. Much later on, when Smolker had moved on to other projects, Connor added underwater hydrophones, GPS mapping, and, most recently, drones to that toolbox.[5]

In time, they were able to determine the gender of their dolphins, which is no easy task. Fortunately, when Smolker went out in the dinghy, she discovered that the dolphins in Shark Bay were swimming alongside the boat. They were, Connor says, "putting their genitals in her face." After hundreds of hours on the water, Connor and Smolker began to create a catalog of photos of all the dolphins and were able to recognize not just the sex of an individual, but also its identity, using the shapes of scars, often from shark bites, on the fin and body: "It's like every time they breathe," Connor says, "they have a little individual recognition flag." This record of dolphin mug shots continues to grow to this day, and rap sheets on more than a thousand animals have been compiled. Shark Bay dolphins live well into their forties, and Connor and his colleagues are still out there gathering data on some of the same animals that were there at the start of it all.

Early on, Connor was collecting data on anything and everything dolphin: aggression and power, mating behavior, synchronized group movement, above-water vocalizations, and "petting," an affiliative behavior in which dolphins touch fins. Still, he didn't really know what *questions* he would focus on for his dissertation. All he knew was that he was fascinated by the system and that it was a potential treasure trove for studying social behavior, including power, in an animal that seemed to fascinate everyone who encountered it.

By 1986, by which time Connor was completely at home with many of the dolphins in the bay, he noticed what seemed like coalitions being formed between sexually mature males seeking to gain mating opportunities with females. "Richard Wrangham, my co-advisor, came out in 1987 and he was blown away by it," Connor says. "I still remember him being out on the boat. There were three trios in sight . . . and he said, 'What, three coalitions in the same area?' And I said, 'Yep.'"

The more he watched these dolphin coalitions, the more fascinated Connor became with the coalitionary route to power in Shark Bay. In time, he gathered detailed information on the coalitions formed among twenty sexually mature males in the population of three hundred dolphins that he, Smolker, and their colleague Andrew Richards were tracking. They ran their results by the Evolution and Human Behavior Group at the University of Michigan and eventually published them in the *Proceedings of the National Academy of Sciences*.

Their paper described pairs and, on occasion, trios of males forming coalitions that chased and then "herded" a female. As males in a coalition chased a female, they engaged in synchronized swimming and aerial leaps. They charged the female, sometimes slamming into her, biting and hitting her with their tails, and often making "popping" vocalizations in the process. The males then followed behind and to either side of the female, remaining together with her for up to twenty days. Herded females sometimes attempted to escape from a coalition, but their success rate was low (about 25%), in part because members of a coalition didn't chase after them haphazardly, but rather cooperatively, angling off to either side to minimize escape paths. The cost, if any, to females of being herded by males is not completely clear, although females do seem to change their space use patterns when they are sexually receptive and males are in the area. Mating opportunities appear to be the payoff of power for males in coalitions: while it was not possible to see dolphins mating from above, Connor and his team did observe coalition members mounting females, and most often females who were receptive (not pregnant or recently pregnant).

Coalitions defended a female they were herding from other males, both lone males and those in other coalitions, and in some cases were even able to herd a female who had been swimming with another coalition. Connor collected data on fifty-eight herding events, involving nine coalitions, and he found that males who formed coalitions to herd females associated with each other in other contexts as well, suggesting tight bonds between coalition members.

When Connor finished his PhD work at Michigan in 1990, the

non-field component of the dolphin project moved to Cambridge, Massachusetts. His PhD co-advisor, Wrangham, had recently taken a position at Harvard, and, still blown away by the coalitions he had witnessed in Shark Bay, he arranged funding that allowed Connor to continue delving into power and coalitions among the dolphins. Year after year, Connor and/or one of his colleagues has been back on Shark Bay, keeping tabs on those coalitions. They now know that coalition members in Shark Bay tend to be genetic relatives—though in other dolphin populations this is not the case—and that some Shark Bay coalitions remain a united front for over twenty years, a figure unheard of outside of primates, and rare even for them.[6]

The ways in which coalitions are used to garner power in Shark Bay seem almost limitless. "I have totally committed my life to studying these alliances," Connor says.

A relatively new mathematical tool suggests that the skills that allow dolphin coalition members to work as a team in the quest for power are honed very early on. About fifteen years ago, animal behaviorists got serious about understanding how social networks—complicated webs that connect social beings to one another—work. To do this, they borrowed the mathematical and computational techniques that were used to build platforms like Twitter and Facebook and modified them to study nonhumans. Social network analysis identifies, among many other things, the "keystone" individuals: those who are deeply connected to others around them and whose removal disrupts the network. Social network analysis also measures something called "eigenvector centrality," which is a gauge of not only how many connections an individual in a network has, but also how many connections its connections have.

Social networks in animals can be simple, involving only a few animals, with information flowing in clear, direct paths between individuals. Or they can be far more complicated, with many individuals embedded in multiple overlapping networks encompassing sub-networks. In both simple and complex animal social networks, interactions between network members have important implications for survival and reproduction. The speed with which information about food, predators, and mates travels in a group depends on social network structure. And, most importantly for our purposes, social networks tell us about power structure.

Dolphin calves in Shark Bay typically stay with their mothers until they are weaned at age three or four. During those first few years, they associate with, and form bonds with, other calves. When Margaret Stanton and Janet Mann examined social networks in the calves, something about eigenvector centrality jumped out at them. Tracking young from birth to age ten, they found that male calves with higher eigenvector centrality were more likely to survive their first decade. Though their social network analysis did not include coalition behavior—males don't enter into coalitions until they are sexually mature many years later—Stanton and Mann argue that learning to form social bonds early on should serve young male dolphins well as they mature in a complex social milieu in which the coalitionary path to power becomes more important over time.

The herding behavior of dolphin coalitions in Shark Bay provides some indirect evidence of greater reproductive success for coalition members, and the social network analysis on bonds suggests yet another likely benefit related to the coalitionary path to power. But as we discussed earlier, whenever possible, what animal behaviorists love to see is direct evidence that a path to power leads to measurable reproductive success. For that, we turn to Claudia Feh's work with Camargue horses (*Equus caballus*) living in the countryside of the South of France.

"Once I was on a holiday in the Camargue [region of France]," Claudia Feh says of a trip she took in the mid-1960s. "I was driving around... looking at the horses in huge pastures.... I looked at horses playing for the first time... and I thought, 'OK, this is the place where I want to live.'" And wild horses were what she wanted to work with. In 1971 Feh took a position doing some ornithological work at a field station in the Camargue region, but gleefully adds, "Of course, I was also looking at horses everywhere." A few years later, the field station initiated a project on Camargue horses, thought to be one of the oldest horse breeds in the world, releasing fourteen animals onto 300 hectares that the field station managed. Feh joined the project as a field assistant, while at the same time working toward her graduate degree in biology.

The study played havoc with her circadian clock, as the work required 48-hour observation shifts, during which she took notes about everything related to the horses and their environment. She'd get up at 4:00 a.m., jump on her little motorbike, and head out to the pasture where the Camargue horses roamed. In the middle of that large pasture was a high-tension electric pole posted with "Danger" and "Stay Away" signs. But Feh thought that pole was the perfect spot from which to watch the horses, so not only did she not stay away, but she built an illegal platform 15 meters up the pole. She'd sit there, binoculars in one hand and a tape recorder in the other, watching one of the horses for up to an hour, then another, and so on, for stretches of three hours at a time. The herd had grown to ninety-four by that time, and despite the fact that all adult Camargue horses are a fairly unadorned grayish-white color, Feh says, "If you see them every day, you know your animals" from the shape of the mane, the nose, and other characteristics.

The aim of that study was not to examine social behavior per se, but rather to better understand the horses' impact on the environment and the way the environment affected the horses. But Feh was interested in behavior, and she saw the opportunity to do deep behavioral observations along with everything the study required. "I spent seven years doing nothing but watching the Camargue horses," she says, estimating that she spent nearly five thousand hours making systematic observations. She also immersed herself in the ethological literature as she began thinking about what the horses were showing her. One thing that struck Feh was that stallions seemed to form long-lasting friendships, and that those friendships paid off, though exactly how she was not certain. She began reading more about coalitions and alliances. "I did not set out to study alliances," Feh recalls, but soon she began to see male friendships as examples of a coalitionary path to power.

After more than a decade of observations (1976–87), Feh dug into her data and her notes to better understand whether coalitions were forming, who was in them, what they entailed, and what their benefits to members were, if any. What she learned was that the stallions use one of three reproductive strategies: guarding a group of females, forming a coalition with another male to jointly guard a group of females, or living in "bachelor herds" and attempting to mate with females in groups that other males are guarding. Most males attempt to guard a group but fail, turning to one of the other reproductive strategies.

Feh focused on thirteen stallions, three of whom were able to guard a group of females, which they did for many years. The other ten stallions, who tended toward the bottom half of the dominance hierarchy in the entire herd, entered into two-horse coalitions. Coalition members were about the same age, and there was no tendency to preferentially form coalitions with genetic relatives. Coalition partners, who often remained a united front for many years, tended to be found near each other year-round and were often involved in intense bouts of mutual grooming. Within a coalition, one stallion was dominant to the other. When another stallion approached a coalition and the mares they were guarding, either the coalition jointly attacked the intruder, or the subordinate member of the coalition confronted the intruder, rearing back on his hind legs and biting or kicking it if it did not back away, while the dominant member moved the mares they were guarding away.

These coalitions paid, and for all parties involved. Foals born to females guarded by members of a coalition had lower neonatal mortality than foals born to females guarded by a lone male. Feh also found that because the subordinate, less powerful member of the coalition often took the lead in fending off intruder stallions, he also assumed higher risks of injury, and probably expended more energy, than his partner. One payoff to the subordinate was that, relative to males in bachelor herds, he sired more offspring. And when Feh looked at cases in which a stallion had been part of a coalition and then, years later, switched to guarding his own group, his reproductive success after switching strategies was no worse than that of a male who had always guarded a group. Surprisingly, this was just as true for former subordinates as for former dominants in a coalition.[7]

Feh tells of seeing the births of many of the stallions in the coalitions she studied. Watching them play together when they were young, she says, made her realize "how important affiliative behavior was" and how critical the ability to form social bonds was to power structure in Camargue horses. Today Feh is retired, but her love for her horses is not: on the fields adjacent to her home, twenty-eight Camargue horses, fifth-generation descendants from the fourteen original animals that started it all, frolic, form friendships, and vie for power.

In both bottlenose dolphins and Camargue horses, once coalitions are formed, they tend to be stable and long-lasting, and their members don't have to constantly solicit aid from their coalition partners. Not so for male olive baboons. They, too, form coalitions to garner power used to find female mating partners, but the baboon coalition landscape is one that demands soliciting aid from others for specific, quite dangerous tasks.

Olive baboons (*Papio anubis*), named after the Egyptian jackal-faced god Anubis, would eventually get Craig Packer onto the pages of the *New York Times*, though that was about the furthest thing from his mind when, as an undergraduate, he signed on to help study them in Tanzania. Packer was a premed student at Stanford University and had no idea how to meet the Study Abroad requirement. "I was vaguely thinking I would go to England, so I would not have to learn a language," he says. That changed one day when Paul Ehrlich, an environmental biologist well known at the time for his book *The Population Bomb*, started showing slides from his trip to East Africa in a course Packer was taking from him. "He showed a picture of a zebra," Packer recalls, "and then said, in typical Paul Ehrlich fashion, 'Well, anyone who wants to see a zebra in the wild better go soon, because they'll all be extinct.'"[8]

At the end of that lecture, a representative from the Study Abroad program told the students that Stanford had a new program where undergraduates could go to the Gombe Reserve in Tanzania and work with Jane Goodall and her chimpanzees. Packer, eager to see a zebra before the species disappeared, signed up. In fact, it turned out there were two options at Gombe: working with the chimpanzees or helping out on a new project on olive baboons. "Being strategic," says Packer, "I thought fewer people were going to apply to work with the baboons and I only really want to go over to see a zebra anyway, so I will sign up for the baboons. And I was one of the first pair to go over and work on the baboons."

During his stint at Gombe, from May through December 1972, Packer got the lay of the land. Aside from naming the animals, his task was to follow the males and observe their behavior. Each morning, with pencil and paper and a check sheet he'd made up, he would head out. The work was difficult and taxing, but both physically and intellectually rewarding. "Gombe is rugged topography," he says. "The baboons might climb up on the hills . . . [a] thousand feet up from the lake. . . . It is quite invigorating . . . and it is very intimate. . . . You can watch them from five meters away."

Researchers had been studying various species of baboons for many years when the olive baboon project at Gombe began. That work had focused primarily on social behavior in females and had found that genetic relatedness structured key social relationships. But Packer, tracking the male olive baboons, who weigh in at about 50 pounds and are about 2.5 feet tall when standing erect, was struck by the fact that "male relationships were very tense . . . very competitive." He was interested enough in the olive baboon social system to enter a PhD program at the University of Sussex, where he built a dissertation project that focused on male dispersal between troops and whether that dispersal reduced inbreeding. Though dispersal and inbreeding, not aggression, were the focus of his dissertation, during his second stint at Gombe, from June 1974 through May 1975, he could not help but notice the coalitions that males were forming, often to get access to reproductively receptive females.

Female baboons, as Packer puts it, "wear their hearts on their butts." Early in a female's estrus (period of receptivity to mating), that area of her body swells dramatically. Packer was surprised that the first males attempting copulations with estrous females tended to be subordinate individuals. Soon he realized that males showed an odd sort of respect for whoever started consorting with a female first. "A male can be a piddling, unimpressive individual," he notes, "but if he is consorting with a female, everybody gets out of the way . . . [because] by being a consorting male you are showing your willingness to be really aggressive in protecting access to that female." Subordinates rely on this deference when females first go into estrus, but soon, Packer saw, "the big boys [dominant, larger individuals] come in and sometimes take females from such males." But then, Packer asked himself, what's a subordinate male to do? His answer was, "You get these coalitions. Two against one. That'll go . . . if [a dominant male] has someone coming in and confronting him and [another] guy who will bite you in the butt . . . you are out of commission. You cannot afford to deal with two against one if they are coordinated."

When he returned to Sussex from Gombe, Packer went to talk with John Maynard Smith, one of his PhD committee members and one of the world's foremost evolutionary biologists. When he went into the office, Maynard Smith was chatting with Ric Charnov, an up-and-coming star in the field, who was visiting from the University of Utah. "They said, 'What's the most amazing thing you found?'" Packer recalls, and, naturally enough, he assumed they would primarily be interested in inbreeding avoidance, the topic of his dissertation work, so that's where he started. "They said, 'Yeah, yeah, so what, that's obvious, of course they will do that. Anything unexpected? Puzzling?'" So Packer regaled them with tales of male coalitions. He added that it wasn't just that two subordinate males formed a coalition to get access to a reproductive female, but that "the guy who asks for help gets it, but then another time the other guy might ask him for help." That caught their attention.

The idea that animals exchange acts of helping was all the rage in animal behavior then, as only a few years earlier, Robert Trivers had published a paper, "The Evolution of Reciprocal Altruism," outlining when natural selection might favor such behavior. Trivers predicted that such exchanges would be especially likely in long-lived species with dominance hierarchies, and noted that "aid in combat" might be one venue where reciprocity occurred. Olive baboons were just the sort of creatures Trivers had in mind. No study had yet systematically tested Trivers's model, and Maynard Smith and Charnov were excited that Packer might be able to do just that in the context of power and coalition formation. And naturally, that got Packer motivated to focus on what role reciprocity might play in the baboon power structure. He went to his field notes from the 1,100 hours of detailed observations and pieced together what was happening.

What Packer discovered was that when a coalition tried to overthrow a dominant male who was consorting with a female, the attempted coup started when one subordinate male recruited aid from another using a series of stereotypical behaviors, which involved shifting his gaze rapidly, over and over, between the individual he was recruiting and the powerful individual that he was targeting. More than 75% of these bouts of solicitation led to coalitions being formed. And olive baboon coalitions paid off. In about a third of the cases, a coalition succeeded in breaking up the consortship between the dominant male and the female. In each and every one of these cases, the enlisting male, who had recruited an ally, formed a consortship with the female.

Joining a coalition to oust a dominant male is a dangerous business, as injury is always a real possibility in these interactions. Packer reasoned that if it was the male who did the recruiting who always ended up consorting with the female, there must be some compensating benefit to the enlistee, otherwise natural selection would not favor responding positively to solicitations. When he had been in Maynard Smith's office with Charnov, Packer had told them that "it was like they [the males in a coalition] were taking turns." His detailed analysis, published years later, suggested that might indeed be happening. Packer found a correlation between how often a male joined coalitions and how often he was successful in enlisting partners when he tried. That pattern was consistent with taking turns, but not definitive proof of it, so Packer next examined coalition partner choice in eighteen males. For each baboon, he looked at who that male most often tried to enlist for an overthrow of the powerful, who Packer dubbed that male's "favorite partner." What he found was that in almost every case, if male 1 was male 2's favorite partner to enlist, then male 2 enlisted male 1 more often than he enlisted other males. This finding strongly suggested that reciprocity is part and parcel of the coalitionary path to power in olive baboons.[9]

That subordinate, less powerful baboons are the ones soliciting partners and forming coalitions is in line with the current, relatively scant, theoretical literature on the evolution of animal coalitions. Ethologists have been gathering empirical evidence on coalitions for the last fifty years, but theory has lagged behind and continues to do so. Two theoreticians, Mike Mesterton-Gibbons and Tom Sherratt, have been attempting to remedy that situation. To do that, they have built a game theory model to understand the conditions that might favor coalition formation. To keep their model as simple as possible—and if you looked at the math, you might take issue with calling the model simple in any sense—they begin by considering groups of three individuals, such that a coalition might (or might not) form between two individuals against the third.

In the model, individuals are assigned strength values—which might, for example, map onto size—from one of various distributions of strength values that Mesterton-Gibbons and Sherratt created. In some distributions they built in a great deal of variance in strength values, with very powerful and very weak individuals; in other distributions, the variance was smaller, such that individuals still differed in fighting ability, but not dramatically so. The model also considered different ways that strength, both that of an individual and that of any coalition it was part of, mapped onto the probability of getting some resource. In some cases, strength was a good predictor of victory; in other cases, it was a less reliable indicator. As with any mathematical model, this one makes a number of assumptions that act as a starting point for the evolution of coalitionary power. Each individual in a triad is assumed to know its own strength, but not the strength of the other two players. In addition, the model assumes there is a cost for joining a coalition and a cost for fighting, should there be a fight.

Mesterton-Gibbons and Sherratt searched for, among other things, the specific conditions that favored coalition formation in the quest for power. One of their findings is fairly intuitive: when the cost of joining a coalition—for example, the risk of injury in joining a baboon coalition—is low, coalitions are more likely to evolve. More interesting is the finding that coalitions are more likely to form when there is a lot of variation in strength values—with some individuals having very high strength values and some exhibiting very low values. Most relevant to the olive baboon study is the model's prediction that under certain conditions, such as when there is high variance in fighting ability, stronger, dominant animals are predicted to go it alone, but weaker subordinates should try to form coalitions, just as the olive baboons at Gombe do.[10]

No animal coalitions, primate or otherwise, are more famous than those formed by the chimpanzees at Arnhem Zoo, brought to the public's attention in Frans de Waal's best-selling book *Chimpanzee Politics*. The wheeling and dealing detailed in that book took place between 1975 and 1981 among chimps in a large colony housed at Arnhem, about an hour's drive southeast of Amsterdam. The brain-child of Anton von Hooff, the large enclosure housing the chimps had been opened in 1971, in a ceremony led by popular ethologist Desmond Morris, surrounded, as de Waal writes in reference to the title of one of Morris's books, by "other impeccably dressed Naked Apes."[11]

In the early years of de Waal's study, the Arnhem colony had four adult males, nine adult females, four adolescent females, six juveniles, and six even younger individuals. The public couldn't tell one chimp from another. All they saw, says de Waal, was "a bunch of black chimps running around. But for us, all of that had content, as we knew the individuals." During the winter, the chimps were housed indoors, but from mid-April to late November, they roamed freely around a 1-hectare outdoor enclosure. De Waal and the students on his team would either use binoculars to observe the chimpanzees from his office overlooking the outdoor enclosure or, on occasion, schlep a huge video camera, which took three people to lug around, out to the moat that overlooks the enclosure, ultimately collecting data on thousands of hours of social behavior.

In 1975, de Waal's first year working as a postdoctoral fellow with the chimps at Arnhem, the chimps were being rather cordial to one another. "When I arrived at Arnhem it was a very quiet time," de Waal recalls. "There was not much going on . . . there was barely any aggression in the colony. And that gave me a chance for a year or so to get to know the chimps very well." That ended when one of the males attempted a coup to grasp the reins of power, at which time, for de Waal, "all the politics started to come [together]." And it was also when he started paying much closer attention to the role of coalitions and how they affected the chimps' power structure.

De Waal saw coalitions among males and also among females, but there were striking differences between the sexes. Male coalitions formed in the context of aggressive interactions and were relatively disconnected from the social interactions between members outside of those circumstances. De Waal has proposed that such coalitions are all about status—that is, that males form coalitions when it increases their chance of rising in the dominance hierarchy. "A male who becomes the alpha male has to have supporters . . . he cannot do that on his own," says de Waal. "He needs to keep [his] supporters happy."

The bonds between females that formed coalitions during conflicts were markedly different. They were stable and long-lasting, and females most often formed coalitions with genetic relatives and "friends" (those with whom they interacted prosocially outside the context of the coalition). "There is a certain level of loyalty between [female] friends," de Waal notes, "and in that sense female coalitions may be more about protecting those close to you than rising in the hierarchy."[12]

Every chimp in the Arnhem population was paying close attention to the complex power dynamics that were constantly playing out. One morning there was a big fight between two males, and that same afternoon de Waal saw "pandemonium . . . [the chimps] were all hooting and yelling and embracing each other and I didn't know what was going on." Later that day it struck him: "The same two chimps that were at the center of that whole event in the afternoon were the same two chimps who had the big fight in the morning. I realized they probably reconciled and that the whole group was excited by it."

Over the years, de Waal has made it clear that ethologists need to be cautious about extrapolating his results to chimpanzee populations in the wild. But he has become more confident that some aspects of coalitions and power struggles at Arnhem are representative of more general patterns in natural populations—at least for males. De Waal's friend and colleague, Japanese primatologist Toshisada Nishida, had long been encouraging him to visit and spend time with the chimpanzee troops that Nishida studied in the Mahale Mountains of Tanzania. Twenty years after *Chimpanzee Politics* was published, de Waal finally took him up on the offer.

What de Waal saw and heard at Mahale suggested that "the males do very much the same thing, they have more space, and a lot is done by sound, but the dynamics of having a coalition partner and keeping him happy [are the same]." One difference, though, jumped out at him: at Arnhem, males did not have to deal with threats from others in neighboring groups. In the wild, they did, and intergroup conflict could be intense.

Life for females at Mahale, particularly as it related to coalitions and power, was markedly different from what de Waal had documented at Arnhem. "The female coalition is more powerful in captivity," he learned at Mahale, "because the females are not spread out over the forest . . . and [so] in captivity the females have a high level of solidarity and their power bloc is more important."

About 750 miles east of the Mahale Mountains, at the Wamba Field Station in the Democratic Republic of the Congo, Nahoko Tokuyama has seen something very different when it comes to power and female coalitions. While studying bonobos (*Pan paniscus*), chimps' closest evolutionary relative, she discovered that female coalitions in the wild are not just real, but potent.

The Wamba Field Station, named after the small village nearby, is home to the longest-running study of bonobos in the wild: forty-seven years and counting. Between 2012 and 2015, Tokuyama and her team collected almost two thousand hours of observations on the dynamics of coalitionary behavior in bonobo Group P. She'd get up at 4:00 a.m., walk an hour and a half to get to the general area where her bonobo group lived, and then follow them with the aid of local trackers. Even with the help of the trackers, this was no easy task, as bonobo groups are constantly on the move, building new nests each place they go.

Group P usually had about twenty-five bonobos. They grew used to having Tokuyama nearby, so she could get as close as about 6 feet, if needed, but she usually kept about 20 feet away, both for safety reasons and to get a wider view of interactions in the group. She would watch each bonobo for five minutes, and then move to another animal, cycling through the group many times a day. Most often Tokuyama was taking notes on a writing pad she always had handy, but on occasion, she'd videotape the animals. Aggression took many forms, ranging from vocal threats to charges, chases, grabs, kicks, and on occasion severe beatings by the aggressor, as well as avoidance behavior, fleeing, grimacing, and screaming by the potential victim. At the end of the day, Tokuyama would head back to the field station, where inevitably people from the local village would come by and pepper her with questions about bonobos, life in Japan, or whatever was on their minds.

Male bonobos are about 25% larger than females, and female coalitions are about protecting one another from male aggression. Of the 108 coalitions formed by females, most involved two or three females, and when they acted as a power bloc, it was most often to threaten, chase, or attack a male that was harassing one of their members. It usually worked: males withdrew and left all coalition members safe and sound about 70% of the time, a figure much greater than when a single female, with no coalition partner nearby, was harassed by a male.

On occasion, females in coalitions truly flexed their muscles. In 2015 Tokuyama saw four females attack an alpha male. That male was part of a group of four males who were harassing an estrous female, when out of nowhere, her three coalition partners came swooping in to her aid. The coalition attacked the alpha mercilessly, and he barely escaped with his life. "We didn't see him for three weeks, and when he came back he was no longer the alpha male, but a low-ranking male that was actually terrified of females," Tokuyama says. "It was a very impressive event."[13]

Assessment strategies, bystander effects, audience effects, and coalitions, be they in bonobos, chimps, baboons, horses, dolphins, hyenas, or any number of other species, show the complexity of the social landscape that must be navigated in the quest for power. In the next chapter, that landscape gets even more complex, for another of its dimensions is the tendency of those currently in power to do everything they can to stop others from following the path to the top.

6   Cement the Hold

Power does not corrupt. Fear corrupts...
perhaps the fear of a loss of power.
JOHN STEINBECK, *The Short Reign of Pippin IV*

In 1973 Steve Emlen, a young professor of ecology and behavior
at Cornell University, was due for his first sabbatical. He intended
to make good use of it: "I wanted to change the world," Emlen recalls, "and work in a complex system." Though he had worked with
birds for many years, Emlen had never seen a white-fronted bee-
eater (*Merops bullockoides*), but his friend and colleague, Scottish
ethologist Hillary Fry, had studied a closely related species in Nige-
ria, and Fry convinced Emlen that white-fronted bee-eaters might
fit the bill for that change-the-world system he sought. Soon Emlen
and Natalie Demong set off to Gilgil, Kenya, located about 180 kilometers northwest of Nairobi, making a quick stop on the way in
Scotland to visit Fry, who, Emlen notes, "was amazingly helpful and
saved me from making gobs of mistakes."

Emlen found "bee-eater colonies all over the place" at Gilgil.
Gorgeous birds with a mixture of green, red, yellow, blue, black,
and white plumage, white-fronted bee-eaters live in colonies found
on cliffsides, where they dig nests about a yard deep into the face
of the cliff. "We started with old-fashioned stuff: marking birds,
taking blood samples," says Emlen, "and then observation, obser-

vation, observation." During that first visit to Gilgil, when Emlen and Demong were working on a single colony that was located on a private farm, things got off to a very rocky start. They thought of the birds not only as beautiful creatures, but as a potential gold mine for studying animal behavior, including power dynamics; however, the locals saw them in a very different light. "We were getting great data. Then, at one point, a number of [locals] came wandering by and just wondered what we were doing," Emlen says, "and being exuberant, I showed them. . . . I thought I was sharing great knowledge with them." What Emlen didn't stop and think about at the time was that what you do with knowledge some-times depends on how hungry you and your family are. "We came back a few days later," Emlen continues, "and all the birds had been snared. The colony had been eaten." From then on, he focused only on colonies that were within protected areas of Lake Naguru National Park.

Soon Emlen was receiving funding from the National Geographic Society and was able to free up additional time in Kenya using a Guggenheim Fellowship he received. Next came money from the National Science Foundation and other agencies, which eventually allowed him to bring in Peter Wrege to head the project and to hire a team, many of whom are Kenyan, who would be on the ground every day studying the bee-eaters.

In time, Emlen and his team mapped colonies scattered across the cliffs in the park and marked all the birds in each colony. "Each nest entrance has a number," says Emlen, who has many oversize photos of the cliffsides of Gilgil showing dozens of nest entrances dug into them. "It's like each hole is an apartment and each one gets an address," he continues, "and you are building up information on who is going in and out of what address."

Each apartment is home to an extended family of bee-eaters, most often made up of a dominant breeding pair, their offspring, and other, more distant genetic relatives. The nonbreeding occu-pants may act as helpers at the nest who feed and take care of the chicks that the breeding pair has produced during the latest breed-ing cycle. These apartments and the neighborhoods around them,

Emlen came to learn, are host to an endless series of power skirmishes, including between fathers and sons.

Emlen and his crew have blinds set up near the cliffs, and one or
more team members sit behind them each day, gathering reams of
data on all things bee-eater. "These guys [the bee-eaters], they are
lazy. . . . You don't have to be there at six in the morning," Emlen says,
describing an average day. "They come out and they are quiet and
huddle together in little groups, which gives you hints of the social
system." Then all the birds leave the area to forage, searching for the
insects that make up their diet. They return to the cliffs after a few
hours, and then head out again, searching for bees and the like.

During the breeding season, on every third day, while the birds
are off foraging, Emlen's team peeks into their nests, using a long
scope with a light at one end, to see how many eggs or chicks are in
there. "You can go in as long as four feet deep," he says, ". . . and get
right over the eggs without touching them. We also had a sort of
'picker-upper,' for when the young got big enough and we wanted to
bring them out and weigh them." On those days, and on every other
day, team members, tape recorders in hand, were assigned an apartment to watch through a telescope once the birds returned from
their foraging. They'd note who was coming in and who was going
out, whether they had food in their beaks, who was fighting whom,
and more. "It's like Hitchcock's *Rear Window*," Emlen says, "you are
watching all these apartment houses, except there is so much more
going on."

As the data began to come in, two things became obvious: First,
helpers at the nest were usually assisting a breeding pair that were
their genetic relatives—most often, their parents. Second, their
efforts had a huge effect on the survival rate of that breeding pair's
chicks. Because nests are dug a yard or more into the cliffside, bee-
eaters are relatively safe from the reptilian and mammalian predators that often cause devastating losses to other bird colonies. Instead of nest defense, the key resource that that helpers provide is
food for the chicks, which is especially important in a system where
starvation is far and away the leading cause of chick mortality.
Having one adult helper assist a breeding pair almost doubles the

productivity of that pair. "I don't think I know of any other cooperative breeder," Emlen says, "where the effect of having a helper help the parent is of such magnitude for the survival of the young."

Most helpers are males that either have never attempted to find a mate and produce chicks on their own nest, or have tried and failed. But some helpers find a mate, secure a nest of their own, and seem to be on the way to success, until an unusual power struggle unfolds. Emlen and his team uncovered that struggle while they were gathering data on the rate at which a male would feed his mate on the days just before she was going to lay an egg. What they saw was that oftentimes, when a male was feeding his mate, another male would fly over and interrupt the feeding. When Emlen consulted the genealogies the team had amassed over the years, he discovered that it was almost always the father of the male doing the feeding who was doing the interrupting. Sometimes brothers or grandfathers tried this as well, but fathers, who are almost always dominant to sons, were the most common culprits.

This was not what Emlen was expecting. As a general rule, natural selection favors helping genetic relatives, particularly close relatives like offspring, not impeding their reproductive success. And, indeed, Emlen knew that better than almost anyone: by the time he saw what those bee-eater fathers were doing, he was already one of the world's experts on how kinship structures cooperation and altruism. What was going on with this father-son power conflict?

When Emlen and Wrege analyzed data from videotapes, they found fathers repeatedly chasing their sons, interfering with their sons' courtship feedings, blocking access to their sons' nests, emitting "begging" vocalizations to redirect food away from their sons' mates, or some combination of these four tactics. In about 75% of such father-son interactions, the son abandoned his own nest and went back to help at the nest of his father.

Emlen and Wrege wanted to know *why*: Why do fathers exert power in such ways? And why don't sons mount a more vigorous attempt to stop them? It boils down to this: *If* a father doesn't harass his son, and *if* his son successfully produces offspring of his own, then the father has new grand-offspring. If a father does harass, and if his son then abandons his nest and helps him, that father will raise more of his own offspring. But individuals are twice as closely related to their own offspring as they are to their grand-offspring, and after a bit of genetic accounting, Emlen and Wrege showed that natural selection, under certain conditions, favors fathers that intervene. For a son, the calculations are different: because his father is more powerful, if a son fights back, there is, at least in principle, a risk of injury. But, even putting aside that risk, if the son stays at his own nest and breeds, he may produce offspring; if he returns to help his father, his help will probably result in the survival of siblings that would not have survived otherwise. Evolutionary biologists have shown that individuals are equally related to their offspring and to their siblings, so there is little selection pressure favoring sons who try to stop their fathers from interfering. Powerful fathers leverage this asymmetry, producing the power dynamics that at first were so puzzling on the cliffs of Gilgil.[1]

To maintain their reign of power, animals have a wide array of tools at their disposal to cement their hold over others—the bee-eaters are just one example. When it's in their best interest, free-living fallow deer in the parks of Dublin break up fights between others to keep them in their place; dominant pigtailed macaques police the behavior of other group members; powerful gelada baboons act in ways that reduce group-level aggression; and superb fairy wrens in Australia, as well as cichlids in Lake Tanganyika, force those lower in the hierarchy to pay a sort of rent to reap the benefits of living under their rule.

How all these strategies operate to maintain power—indeed, why they work at all—are active areas of research in animal behavior. Teams of researchers from around the world are digging into why we see these sorts of behaviors here, but not there; in one sex, but not in the other; among kin in some cases, and among strangers in others.

Dubliners love the hundreds of fallow deer (*Dama dama*) that roam free in Phoenix Park. "The deer have been there for hundreds of years," animal behaviorist Domhnall Jennings says, "and they have

settled in quite nicely. . . . [People] are quite protective about them."
Jennings has been out there, in the park, taking data on those deer,
with a special focus on their power dynamics, on and off for the last
twenty-five years (and counting).

Just a twenty-minute bus ride from the heart of Dublin, at 1,750
acres, Phoenix Park is about twice as big as New York City's Central Park. The park traces its history back to the 1660s, when it was
a royal hunting ground, and fallow deer were brought in for sport.
In 1747, with its resident deer still in place, the hunting ground was
transformed into a public park and has remained so going on 275
years.

Jennings was keen on studying contests and the evolution of aggression for his PhD, when one day he bumped into Tom Hayden,
who was leading a long-term project on fallow deer in Phoenix Park.
Hayden suggested that perhaps the deer system would work for
what Jennings had in mind. "Kind of one of those things," Jennings
says. "You bump into someone and you end up doing a PhD. . . . [It
was] a whirlwind. . . . One day in September I was wondering what
to do . . . and the next day I was told to go out to Phoenix Park and
collect data."

For his dissertation, Jennings focused on pairwise male aggressive contests, especially during the rut—the mating season—in
mid-October. At the time, there were about 750 deer in the park,
and most of them were tagged for individual recognition; those
that weren't could be identified by antler shape and fur markings.
He would get to the park before sunrise and use a spotting scope to
watch the deer, while jotting down notes. He also carried an old video
camera around with him on occasion: "one of those big VHS jobbies,"
as Jennings describes it, "[and] I had a battery pack strapped to me.
. . . I would be running around Phoenix Park. It is such a large place,
I used to cover miles every day."

Over time, Jennings's team has grown. Connected by radios
and armed with spotting scopes and, later on, the voice recorders
and video cameras on their smartphones, they stay in touch trying to cover as much of the park as possible, noting, among many
other things, when pairwise contests between males occur, how

many females are the vicinity, and more. On occasion, they have to pause to pay the price for doing ethology in a public park. "One guy picked up two branches and was kind of running around like a buck," Jennings recalls, "and I said, 'Don't do that.'" More seriously, he or one of his team sometimes has to stop and lecture passersby not to feed the deer, because that will lead to the deer losing their inhibitions around humans. Once that happens, Jennings is worried that it is only a matter of time before "one of the big males will attack somebody someday."

When Jennings, Hayden, and their colleagues sat down and analyzed the almost two hundred contests they had videotaped in 1996 and 1997, they found that male power dynamics were often quite subtle. "One of [the males] will walk up to another one and just nudge him out of the way," Jennings says. Other times, he "wouldn't even touch him." Some encounters include a parallel walk, in which two males walk slowly, side by side, vocalizing, often with hair erect and antlers raised up high. In red deer, males use information from such parallel walks to determine whether to escalate to more dangerous contest behaviors, but Jennings found that was not the case in his fallow deer.

When power contests do involve significant physical contact, things can get very intense, very quickly. Males lock antlers and try to push their opponent backward. At times, that is enough to settle matters, in which case the males unlock antlers, and one retreats, while the other either lets the loser leave in peace or pursues him for a bit. The most intense and dangerous contest behavior, what Jennings and his colleagues call a jump clash, involves one buck lowering his antlers, and either rearing back on his hind legs and crashing into his opponent or leaving the ground completely (jumping) while crashing into the other buck. This is serious business: on occasion, jump crashes lead to broken antlers and even skull damage.[2]

Pairwise power contests continue to be part of Jennings's research agenda, but a few years after his PhD work, he began to wonder whether he was missing a key component of the power struggles among fallow deer bucks. Pairs don't contest for power in a vacuum, and those around them, particularly those more powerful than the contestants at hand, might have their own stake in contests that play out between others. As Jennings began diving into the literature, he realized that powerful males might have cause to intervene in the contests around them. He recalled seeing interventions on several occasions, and he knew there were probably more on the videotapes, but he had never made them his focus. That, he decided, needed to change, so he began digging back into those tapes and the notes he had gathered during his dissertation project, searching for information on the dynamics of interventions.

What Jennings found was that interventions occurred in about 10% of all pairwise interactions and came in two flavors. "It can be *bam*—it just happens," Jennings says, as seemingly out of the blue, one of the males in an ongoing contest "gets whacked by a [third-party] 120 kilo [264-pound] buck and [ends up] somersaulting through the air." That said, most interventions are less dramatic and less dangerous. "Often two males will be fighting," he continues, "and a third male will just sort of wander over toward them. . . . The pair might start parallel walking with this guy just trudging along after them." At some point, the third party either leaves or it intervenes, often by locking antlers with one of the contestants and engaging in pushing contests or jump clashes, breaking up what had been a pairwise interaction.

Jennings and his colleagues have found that the number of interventions in an area increases with the number of estrous females. But they found no relationship between the number of interventions and the number of matings in a herd on a given day, which raises the question of why males intervene in fights to begin with. To address that question, Jennings tested a model that I had published on winner effects and intervention behavior a few years earlier. Though a winner effect has not been explicitly tested in fallow deer, indirect evidence suggests that one is in play. In my model, interveners do not target an individual in a pair, but rather attempt to break up interactions between that pair. The model suggests that intervention behavior can be favored by natural selection if it has the effect of preventing others from getting on a winning streak and becoming a more serious threat to the intervener. It predicts that individuals at any rank in a hierarchy may engage in interventions, but that interventions by high-ranking individuals should be most common.

Jennings found some support for this model: high-ranking bucks were most likely to intervene, and the higher an individual's rank in the herd hierarchy, the more it intervened. In addition, interveners did not appear to be targeting a specific male in a pair: one male in a pair often got the worst of it in a pushing match or jump clash, but Jennings found no evidence that which male did was a function of size, relative rank in the pair, or prior interactions with the intervener. To break up a fight and prevent a winner effect from kicking in, someone has to be attacked, but who that is seems to be random. The upshot is that neither of the pair wins the contest they had been engaged in, and so any winner effect that might have kicked in, and potentially made one of them more of a threat to the intervener later on, is nipped in the bud. That benefit to the intervener is magnified in fallow deer because males who are part of a contest that ends without a victory, as is the case when a fight is broken up, are more likely to be part of indecisive contests in the future, which further curtail any winner effects ever amassing for the victims of intervention. The intervener also benefits in another way, in that compared with interactions he has with others in the herd, he's more likely to win future interactions with *either* of the males in the pair he intervened upon.

When Jennings dove into the literature, he found that while there was some information on the benefits that a buck gets for intervening, there was virtually nothing on the costs of being the victim of an intervention. That didn't make sense to him: "You are fighting away, putting all this investment into it and then it gets disrupted and so it is basically wasted investment . . . seemed to me a gap that needed to be filled." He filled that gap quite nicely in his paper "Suffering Third-Party Intervention during Fighting Is Associated with Reduced Mating Success in the Fallow Deer."[3]

Intervention as a tool wielded by the powerful plays out in a different, but equally intriguing, manner in pigtailed macaques.

Plate 1. A coalition of young spotted hyenas attacking a larger, but lower-ranking, individual at the Masai Mara Reserve in Kenya. Photo courtesy of Kate Yoshida.

Plate 2. A power struggle between two giant Australian male cuttlefish for mating opportunities with the smaller female seen in the background. Photo courtesy of Roger Hanlon.

Plate 3. Power struggles between loons on the lakes of Wisconsin can be hyperaggressive and, on occasion, lead to the death of one of the combatants. Photo courtesy of Kevin K. Pepper.

Plate 4. Coalitions are integral to the power structure among the male bottlenose dolphins of Shark Bay, Australia. Coalitions guard females with whom they mate. Pictured here are a second-order alliance made up of two first-order alliances following two females (*top left*); a first-order alliance, a female, and a calf (*center*); and a second-order male alliance of three trios, and two females (*bottom right*). Photo courtesy of Simon Allen.

Plate 5. A dominant male olive baboon (*far right*) was challenged by a pair of older males, but stood strong and backed the pair into the waters of Lake Tanganyika. Photo courtesy of Craig Packer.

Plate 6. Two groups of banded mongooses face off in Queen Elizabeth National Park, Uganda. Moments later, an intense between-group power struggle ensued. Photo courtesy of Harry Marshall/Banded Mongoose Research Project.

Plate 7. A Florida scrub jay family stands guard against intrusions from neighboring groups. Photo courtesy of Reed Bowman.

By the time Jessica Flack started working on power and interventions in pigtailed macaques (*Macaca nemestrina*) at the Yerkes National Primate Research Center, a fair amount about power and aggression in this monkey was known from fieldwork done by Irwin Bernstein and Toru Oi. By the early 1960s, Bernstein had worked extensively with pigtailed macaques in the laboratory, so when he began one of the earliest field studies of these animals in a rainforest along the west coast of Malaysia, near the Bernam River between Perak and Selangor, he assumed he knew enough to at least get the monkeys comfortable having him there watching them. But the macaques did not take well to having a human lurking so close by. And probably for good reason, too, as they were aggressively hunted as food by the local people.

To get the pigtails used to his presence, Bernstein first used a tried-and-true method employed by field primatologists: "On detecting the troop, I remained motionless," he wrote, hoping the monkeys would go about their normal routine. Instead, they fled into the dense rainforest, filled with its giant meranti, cengal, keruing, and merbau hardwood trees. Next, he tried bribing them, provisioning them with food. Again, no luck, probably because the locals may have tried to lure macaques into their traps that way. Then Bernstein got a bit more creative. He purchased a pigtailed macaque from a local market, trained it, and then walked around the forest in sight of the troops of macaques that called that area home, now partnered with a 20-pound shill sitting on his shoulder or walking beside him. That did the trick, and he could finally take to gathering detailed observations of the two troops he wanted to study.

After four hundred hours watching and taking notes on anything and everything macaque, Bernstein began to put together a rudimentary picture of life in a troop. He published papers about when they moved and where, who played with whom, who preferred whom as a mate, their feeding habits, their vocalizations, and, of course, their power structure. The dynamics of that power structure involved chases, threats, and attacks aplenty. In his papers, Bernstein also made passing reference to monkeys who "interfered" in the aggressive interactions of others "without being enlisted to do so." Because he couldn't recognize specific individuals, he was not able to piece together a more detailed picture of power, but he set the stage for Toru Oi to come one step closer to doing just that.[4]

For two years, starting in January 1985, Oi worked with pigtailed macaque troops in West Sumatra, at the base of Mount Kerinci, in a rainforest with an annual rainfall of about 200 inches. The area was a mixture of primary rainforest and secondary growth—such as wild ginger and bananas—on what had once been tea and coffee plantations, but had been long since abandoned by early Dutch colonizers. Oi had it easier than Bernstein getting the pigtailed macaques used to having him nearby, as they took well to the peanuts he provisioned them with, which meant that he could recognize all twenty-six adults and adolescents in the main troop he was watching.

Power plays among the macaques of West Sumatra included chases, slaps, "hold-downs," and bites. Males were dominant to females. Drawing on over two thousand recorded acts of aggression, Oi constructed two dominance hierarchies, one for males and one for females. For the most part, though not in every case, rank in the male and female macaque hierarchies was stable over the course of the study. In addition to recording pairwise interactions, Oi also observed three-way (triadic) interactions, which sometimes involved a third party supporting one of two protagonists or intervening and breaking up fights between them. All of which is to say that the early fieldwork by Bernstein, Oi, and others suggested that power struggles in macaques are complicated affairs.[5]

Starting in the early 2000s, Jessica Flack, a PhD student at the time, along with her mentors Frans de Waal and David Krakauer, ran a series of experiments on power in a group of eighty-four pigtailed macaques at Yerkes. What they knew when they began was that when a macaque is involved in an aggressive interaction with a more dominant individual, it often emits a "bare-teeth display," in which its lips are retracted and its teeth are partially exposed. Pigtails, and other closely related macaque species, have two variants of the bare-teeth display, one accompanied by vocalizations, and the second a silent variant. Work by Flack, de Waal, and others found that when the lower-ranking of two pigtailed macaques uses a bare-teeth display during an aggressive interaction, it is almost always a silent version. After that, the encounter often ends.[6]

As they studied the power dynamics of their troop, Flack and her colleagues observed many an intervention, during which an individual, typically a high-ranking individual, approached a pair in an ongoing dispute. Often just the approach, or some sort of aggressive display, by the dominant would result in the breaking off of that interaction, but on occasion, the intervener would employ low-level physical aggression, again resulting in the pair breaking off their interaction.

Flack was intrigued by all this, and so she and her colleagues watched the troop for more than 150 hours and recorded 447 acts of impartial intervention, where—at least in Flack and her team's eyes—the intervener did not act to aid one of the two participants. Of these, 189 were successful, putting an end to the aggression underway. But intervention was not cost-free. In about 10% of cases, one of the two contestants resisted an intervention, at times with serious acts of aggression, including biting, toward the intervener.

Not everyone intervened. Four individuals at the top of the troop's hierarchy not only were involved in the majority of the successful acts of impartial intervention, but were much less likely than those lower in the hierarchy to have their acts of intervention resisted. This observation got Flack wondering what would happen to the system if she experimentally tinkered with the power structure by removing the individuals most responsible for the interventions.

To disrupt the pigtail power system, Flack and her colleagues imported, and then modified, a tool from molecular genetics. One way geneticists study the relationship between a gene and its function is to run a "knockout" experiment, in which they use genetic engineering techniques to slice out or replace key bits of DNA to inactivate a gene, making it nonfunctional. If a trait becomes nonfunctional, or is simply absent, as a result of the gene knockout, this suggests a strong causal link between that gene and the trait in question.

Flack and her team "behaviorized" this knockout tool. On randomly selected days during a twenty-week study, they monitored the entire troop of pigtails while all members were free to move about in their indoor-outdoor enclosure. Once every two weeks, again on randomly selected days, three of the four monkeys that were responsible for the vast majority of interventions were confined to the indoor enclosure for stretches of ten hours, while the rest of the troop was free to do as it pleased in the adjacent, much larger, outdoor area. Large window boxes lying between the indoor and outdoor enclosures allowed visual and vocal communication, but not physical interaction, between those removed and other troop members. The three pigtails on the inside often sat right next to the window boxes, which prevented them from making any interventions, but at the same time increased the probability that other troop members, who could see and hear them, would continue to act as if these powerful individuals were still troop members.

Effects were immediate and dramatic. None of the macaques in the outdoor enclosure increased their intervention behavior; instead, with the frequent interveners behaviorally knocked out, the rate of aggression among other troop members increased, and the rate of prosocial behaviors such as play, grooming, and post-fight reconciliation decreased. Even more remarkably, during the periods when the interveners were confined, the rest of the troop underwent a kind of balkanization, in which smaller, more homogeneous groups that rarely interacted with outsiders replaced the larger, more diverse social network seen when the interveners were free to manage group-level conflict. As soon as the removed individuals were allowed to rejoin the troop, all these changes disappeared— until the next knockout period, when they quickly reappeared.[7]

The results of the knockout experiment are consistent with the idea that removing the individuals responsible for most interventions led, at least in part, to many of the changes in the power structure that Flack and her colleagues observed. But interpretation of these results is complicated, as Flack and her team understood all too well. Would removing individuals that didn't intervene have led to the same results? A smaller-scale experiment using the same protocol, but removing a low-ranking individual that did not intervene, produced none of the large-scale changes seen in the original experiment. While that smaller experiment involved only a single individual being knocked out, and its results should therefore be taken with caution, they are at least consistent with the idea that removing interveners is what mattered in the original experiment.

Another possible confounding factor in the original experiment was that by removing individuals who intervened, Flack and her team were also removing high-ranking individuals, and perhaps it was rank, not intervention behavior, that was driving the changes. Maybe the changes occurred because all the remaining individuals in the troop were in the process of re-forming the hierarchy. It is possible, after all, that even though the knocked-out individuals were removed for only short periods of time and were in sight and sound range, the remaining individuals acted as though those knocked out were no longer part of the group, despite Flack's attempt to minimize that possibility. This is a tricky alternative hypothesis to test, because to remove interveners, Flack *had* to remove high-ranking pigtails, as they were the ones intervening. So, to address whether individuals were attempting to restructure the hierarchy, rather than responding to the absence of intervention per se, what Flack and her colleagues did was to look for the sorts of interactions known to be associated with unstable hierarchies that are in the process of re-forming. Those interactions include subordinates challenging those above them in rank and encounters in which both parties, rather than just one monkey, are aggressive. The team found no evidence for either.[8]

All in all, then, interventions appear not only to affect power, but to have an effect, both directly and indirectly, on the very nature of pigtailed macaque society.

Interventions affect the power structure rather differently in gelada baboons (*Theropithecus gelada*). Working with a captive population at the NaturZoo in Rhine, Germany, Elisabetta Palagi, Virginia Pallante, and their colleagues found that both high-ranking female and high-ranking male geladas often intervened in an ongoing dispute. But these intervening individuals tended not to be impartial, most often supporting the lower-ranking member of the contesting pair. When the original pairwise dispute was over, the intervener often sat close to the individual it had helped, embracing, playing with, and grooming it. This behavior seemed to have a calming effect on former victims, as evidenced by a decrease in scratching behavior, a sign of anxiety in gelada baboons and other primates. An interesting consequence of these interventions is that they affected power struggles among *other* group members besides the three involved in the intervention. When an intervention supported a low-ranking individual, the aggregate level of aggression among those *not* involved in that intervention decreased, while the converse was found when the intervener supported the higher-ranking member of a pair.[9]

Palagi has recently begun following up her zoo work with field studies of a free-living population of hundreds of gelada baboons living in the Kundi highlands of the Amhara Region of Ethiopia. But she's glad she did the zoo work first: "When you go to the wild, it is very hard to know [beforehand] what it is going to be possible to [see]," Palagi notes, "but if you go to the wild with a clear idea . . . it is easier." She says her that results are still preliminary but encouraging, in that what she has seen while perched on her platform watching, and at times videotaping, the wild baboons resembles what she's seen at the NaturZoo.

The Kundi highlands of Ethiopia are a striking place to study power dynamics. But then so are the Australian National Botanic Gardens, albeit for different reasons.

Spread across the 30 acres of the Australian National Botanic Gardens in Canberra are acacia and eucalyptus trees, orchids galore, and some six thousand plant species — 63,000 individual plants — to see, smell, and touch. But it isn't just plants that the nature lover can soak in. The gardens are also home to a population of superb fairy wrens (*Malurus cyeneus*). These smallish birds are about 6 inches long from the beak to the tip of the tail and weigh a minuscule one-third of an ounce. During the breeding season, in late spring and summer, males have gorgeous iridescent blue plumage set against a backdrop of black and gray feathers and an orange beak. Females and juveniles display drabber brown plumage, though they, too, have lovely orange beaks, and some orange under the eyes as well. Superb fairy wrens' nests are dome shaped, with a side entrance. Raoul Mulder learned that within and around that entrance during the breeding season, power plays out in fascinating ways.[10]

Like white-fronted bee-eaters, superb fairy wrens are cooperative breeders. A dominant male and female produce young. Other sexually mature males temporarily delay dispersing to set up their own territory and find a mate, instead remaining at their natal nest to help the (dominant) breeding pair raise their offspring. Mulder, who was interested in learning more about fairy wren mating and courtship behavior, was immediately struck by one observation: "I started noticing all this sort of weird behavior by males, in particular this petal-carrying behavior they do. Males would fly up to a bush and pick up bright yellow acacia flowers and go and display them to females," Mulder says. "These displays never resulted in [immediate] copulations. It was really strange, why were they doing this?" What he discovered was that petal displays were directed at females who were living on a territory other than the displaying male's, and who were already paired with the male on that territory. Which is to say petal carrying was a courtship display that might eventually lead to what have been dubbed "extra-pair copulations." Indeed, genetic fingerprinting work would soon reveal an extraordinarily high rate of extra-pair copulations in this species: close to 75% of all chicks at a nest are not sired by the dominant male at that nest.

All these extra-pair copulations result in a situation in which helpers at a nest are not closely related to the dominant pair there. Helpers almost never sire any of the young at the nest of the dominant pair, but they are responsible for the vast majority of territorial defense early in the breeding cycle as well as for feeding the chicks at the nest.[11] "In larger groups, even though the dominant male is getting hardly any paternity, he's also doing almost no work at all in terms of feeding of the offspring," Mulder notes. "That led us to questions about how this is all enforced. Why would the helpers keep helping? What's keeping them in line? Why are they dutifully turning up and feeding if they are not closely related?" Mulder already had a good feel for why helpers delayed leaving their natal territories—that was primarily because females and territories were in short supply—but, given that they stay, why do they feed the dominant pair's young and defend the nest?

Mulder hypothesized that when mates and unoccupied suitable territories are scarce, helpers pay "rent" to powerful territory holders in order to enjoy the comfort of a safe spot to live until better times come. If this "pay to stay" analogy, first proposed in the late 1970s by A. J. Gaston, was correct, then the dominant male should respond when that "rent" is not paid. To find out whether that actually happens, an experiment was called for.

Working with animal behaviorist Naomi Langmore, Mulder simulated an act of cheating on the part of a helper by removing him from the territory for twenty-four hours, during which he could do nothing in terms of defense and feeding—in effect, forcing him to default on his rent. Then Mulder and Langmore brought that helper back to the territory again to see how the dominant male responded.

The administrators at the gardens insisted that Mulder and Langmore be done with their work by 9:00 a.m. each morning, so in the evenings they'd plan out which nest to target, and then early the next morning they would set up a net with the hope of catching the helper they had targeted the prior evening. "If it was a lucky day, we'd go out . . . catch the target bird, head back to the university across the road, put [him] into a cage with food and water and leave him in his quiet hotel for the remainder of the day," Mulder says, "and on the next day we'd put him in a bag, go back to the territory, and do the release." He and Langmore, with binoculars and a tape recorder in hand, would then watch (and record) what happened.

Removals were done at three times during the year. In the first removal period, when males had just molted to their iridescent breeding colors but the breeding season had not yet begun, six helpers were removed. "It was the biggest nonevent ever," Mulder says with a smile. "We captured a helper, brought him into captivity, put him back, and it was like he had never been missing. No overt hostility. Nothing." Then again, he knew this was a time of the year when helpers were not defending a nest with eggs or feeding chicks, as there were none to defend or feed.

When they repeated the experiment, first when females were incubating eggs, and again when chicks had hatched, things got much more interesting. After those fourteen removals, the dominant male clearly responded to the helper's failure to pay rent. In many cases, "no sooner had the helper flown out of his bird bag [back] into his territory than the dominant was on his case, relentlessly pursuing him through the territory," says Mulder. "It was quite clear the dominant male . . . had noticed that that helper had been missing. . . . These pursuits would last four or five minutes." On occasion, the dominant male caught the helper and pecked him, though Mulder saw no obvious injuries even then. These pursuits could continue, on and off, over one or two days, but in every case, they eventually stopped, perhaps because the point had been made. The helper was accepted back into the group, and behavioral interactions returned to what they had been before Mulder and Langmore's act of bird-napping.

Mulder and Langmore considered a number of alternatives to the explanation that powerful males were punishing delinquent helpers. Perhaps the dominant male failed to recognize a helper after it had been removed and then returned—in effect, maybe the bird who was a helper was now being treated as a stranger. The evidence says no. Helpers always assimilated back into their group within a few days, but other males from outside the dominant's territory were chased off by the dominant until they left for good. Mulder and Langmore also considered the possibility that dominant males' aggression toward helpers upon their return was simply a matter of their having high levels of testosterone and being generally aggressive. But, no, during that first removal period, when helpers were not punished upon their return, dominant males had their breeding plumage, and though they had not yet started breeding, they had high levels of testosterone. So it wasn't just high testosterone levels that led to the dominant's response. Returning to the questions that drove this work in the first place—Why would the helpers keep helping? What was keeping them in line?—Mulder and Langmore suggest that it's about paying (or not paying) rent: powerful male superb fairy wrens punish helpers who don't provide food and defense at times when they are needed most.[12]

Cichlids pay rent, too. With its icy-blue eyes and blue-tipped fins, *Neolamprologus pulcher*, sometimes called the "Princess of Lake Tanganyika," also has helpers at the nest and uses a variant of the pay-to-stay power system that fairy wrens use. Helpers increase the reproductive success of the territory-holding pair they assist, but, as Michael Taborsky, who studies these fish in the wild and in the lab, discovered, they add new layers of complexity, in that the helpers can be either male or female, large or small, and related or not to the dominant breeding pair on a territory.

When Taborsky began the field component of his PhD on *N. pulcher*, he worked in Burundi, at the northern end of Lake Tanganyika. He and Dominique Limberger, who was doing PhD work on the same species but whose interests leaned more to the physiological than the behavioral, rented a small hut near the lake, owned by a fish-processing plant. Like all PhD students, they were itching to get started, but, as with so much fieldwork, the unexpected slowed them down.

"There was a cholera epidemic when we came to Burundi," Taborsky says, "and we were not allowed to access the lake. We had to negotiate quite a bit with authorities." Cholera was not the only danger. The water was teeming with hippos, and when Taborsky spoke with local fish exporters who were sending thousands of cichlids to pet stores and breeders around the world, they told him the cichlids he wanted to study were in an area where, just weeks earlier, a hippo had killed a diver. Apparently that diver was the fourth person to meet this fate in a relatively short period of time. "They warned us," he said, "dive deep enough where the hippos cannot reach you." That was not always possible, as the prime sites they found for observation were only about 3–5 meters deep, so they were understandably "quite scared, looking around whenever we heard something."

Each day, Taborsky would head out on a boat steered by a local fisherman, put on scuba gear, and spend four to six hours observing *N. pulcher*, jotting down notes with an underwater marker on a PVC board. What he saw straightaway were breeding pairs with about six helpers defending territories, each of which included a hole or cleft in the lakebed that was used for breeding. Helpers assisted in cleaning and fanning the eggs of the breeding pair, digging sand from the area around the breeding shelter, and defending against predators as well as intruding *N. pulcher*.

Taborsky learned two things early on. For one, there were no sub-stantive differences in the way male helpers and female helpers be-haved. For another, sexually mature helpers were not staying at the nest of their parents, nor were unrelated helpers joining new nests, for lack of places where they could live on their own: there were plenty of such places available. Instead, individuals acted as helpers to a dominant pair because it was often too risky to try to live on their own. "We found from experiments in the field," Taborsky says, "[that] you have very little chance of surviving if you are on your own . . . without the protection of the group that defends a shelter." In time, he also learned that when one or both breeders die, sometimes, though relatively rarely, larger helpers, who tend not to be related to the breeding pair, take over and make the territory their own.

Given the limited alternatives available to helpers, Taborsky began wondering whether dominant individuals might not use the situation to their own advantage. "Subordinates get access to the re-sources available in a territory, especially protection in territories, but, of course, [maybe] they have to do something for it . . . a kind of paying for being allowed to stay, of paying rent to reap the re-wards of group membership and being allowed in a safe territory." To better understand if that was the case, he and Ralph Bergmüller ran an experiment at the Institute of Zoology in Bern.

Bergmüller and Taborsky created territories that housed a breed-ing pair and two helpers, one large and one small. In phase 1 of the experiment, all four group members were presented with a threat from an intruder male who was not part of their group. Defensive behaviors toward that intruder, as well as aggressive interactions between group members, were recorded. In phase 2, using a clever combination of partitions, in some trials one of the helpers was placed in a compartment where it could not see the intruder, but all other group members could see that helper as well as the intruder. Then, in phase 3, the entire group was re-formed and was presented with an intruder that everyone could see.

Bergmüller and Taborsky predicted that during phase 2, other group members would compensate for the inactive helper and in-crease their defensive behaviors, and that is indeed what they did, taking up the slack, particularly when the intruder was large. But they also predicted that the helper who had not helped in phase 2 would be punished. Here, the results got interesting. In phase 3, the breeding pair did not increase aggression aimed at the helper who had not seen the intruder, although the pay-to-stay model predicted they would. Instead, evidence supporting the pay-to-stay model came in a different form. Helpers, especially small helpers, who had been prevented from seeing the intruder in phase 2 increased their defensive behaviors in phase 3 (compared with phase 1), possibly preempting any punishment from other group members. More recent work suggests that sometimes, depending on exactly the type of helping behavior and the context in which it occurs, dominant *N. pulcher* individuals do punish helpers who fail to help. Taborsky calls this variation in the tendency to punish those who fail to pay rent "commodity-specific punishment," and its existence suggests a complex pay-to-stay system in which all individuals, the powerful and the not so powerful, respond to multiple environmental and social cues.[13]

White-fronted bee-eaters, fallow deer, pigtailed macaques, gelada baboons, fairy wrens, and cichlids show us that the powerful will do whatever they can to control those below them. Each of these examples was embedded in group life. Indeed, almost every example in every chapter of the book to this point is about the struggle for power in social, group-living animals, and we've touched on various attributes of the groups that these power struggles take place in. But that is a very different matter from where we now turn—power at the group level.

7   Survive the Battles

> All animals are equal. But some animals are more equal
> than others.
> Napoleon the Pig, in George Orwell, *Animal Farm*

When it comes to dolphins in Shark Bay, in addition to what hap-pens *within* coalitions, coalitions often unite and try to exert power over other coalitions, layering complexity on top of complexity. "Second-order" coalitions are born when members of two coalitions join forces to secure new reproductive opportunities or to ensure that the access that one of the coalitions has in place is not lost. These powerful super-coalitions "are stable for decades," Richard Connor says, but then adds with a smile, "Sometimes if there is an opening . . . an old fart may join if he has lost his buddies." If all that isn't enough, second-order coalitions themselves sometimes come together to form, for lack of a better term, third-order mega-coalitions that compete for power with other mega-coalitions.[1]

For banded mongooses living in Uganda, it's different: inter-group power dynamics are a much bloodier affair than for dolphins, coalitions or no coalitions. "Battle lines," as Michael Cant, who has been studying mongooses for twenty-five years, calls them, are not to be crossed. If they are, all hell breaks loose.

Cant first went to South Africa to study mongooses but couldn't find a good site, so he headed over to Queen Elizabeth National Park in Uganda, where Danielle De Luca, a PhD student from University College London, was finishing up her dissertation work on the costs and benefits of group living in banded mongooses. "Danielle showed me how to catch mongooses, immobilize them, basically taught me the ropes," says Cant.[2]

Today Cant, both when he is in the field and when he is at the University of Exeter, leads a team of seven Ugandan assistants, supervised by Francis Mwanguhya, who has been with the project since its inception. Not only do all the team members know all the animals, but some of the banded mongooses appear to know specific members of the research team. "There was one female," Mwanguhya says, who "did not like one of the researchers. Every time he turned up in the group, this particular female made a grumbling noise . . . because this researcher had captured her group and [kept] them in the lab for [many hours before] releasing them." Mongooses can hold a grudge with the best of them: Mwanguhya was beyond surprised when, after his colleague was away for more than a year, he was greeted with the same grunts from the same female mongoose upon his return.

Cant's early work focused on why there was high reproductive skew in groups: why a few dominant females produced young while suppressing reproduction in other females. Soon he began to learn that "mongooses are always puzzling, always surprising, they do everything wrong. . . . They keep us scratching our heads for all sorts of reasons." One of those reasons is that the dominant females in a group all give birth on the same day—indeed, on the very same morning. It happens on different days in different groups, but within groups the synchrony is astonishing. "You go there," as Cant tells it, "and four or five females are massively pregnant and wobbling around. Then the next morning . . . at eleven o'clock they emerge, all slim. We still don't know how that [synchronization] happens."

One of the other things that had him scratching his head, and ultimately led him to study between-group power scrums, was the periodic violent eviction of a subset of females from groups that were growing quickly. One day you see "peace and harmony and love and nibbling each other's necks, and then you go back the next day and all hell has broken loose" is how Cant describes what unfolds during these mass evictions. "Some individuals, for reasons we don't know, have been labeled evictees and they just get relentlessly attacked by everyone. . . . It is primarily females attacking females, but once an animal has been marked for eviction, even the little pups will come along and join in the kicking. There is a mob psychology to it."[3]

It was following one of these mass evictions during Cant's PhD work that something completely unexpected happened. Ten females were evicted from a group and approached a large group composed of twenty males and eight females. That was not the unexpected part. Mongooses live in groups. Period. And so after an eviction, the most important thing to do is to get into or form a new group. "I went there in the morning and the big group was fighting against the females," Cant says, "driving them off." That, too, was not so surprising. But "then I went off and got some lunch, and when I came back in the afternoon, nine males had left the [large] group, joined the females, and were now fighting against their old group." That was surprising. What started as an eviction had resulted in a full-fledged between-group power struggle.

As memorable as that event was, Cant came to learn that it's not the typical way in which power struggles between groups unfold. Banded mongooses breed four times a year (with January and February as their two down months), and, sans a forced eviction, individuals don't usually disperse from groups. That leads to genetic relatedness within groups building up over generations, which, in principle, could lead to inbreeding—except that females have found a solution. They go searching for mating opportunities in neighboring groups. When they do, they are often followed by males from their own group, which has the effect of inciting fights between groups. "In the middle of the battle," Cant says, "females go and mate with males from the other group."

These between-group power clashes are a sight to behold. "When they are fighting, they are like single organisms," Cant says, ". . . writhing balls of fur with claws, and they have these battle lines they draw . . . chasing in and out of bushes . . . screaming. . . . Sometimes you suspect that animals don't even know if they ended up on the wrong side." It's such chaos that even when Cant and his whole team are there observing, they can't tell who's who. They hope the flying mechanical drones that they are starting to use, when tied to deep-learning AI programs, may help on that front. At the end of these clashes, which can last many minutes, there are often many casualties, including deaths, among males. Females almost always walk away unscathed—which is not to say that between-group power struggles are cost-free for females. When Cant and his team looked at pup survival rates, they found that pups at the den were less likely to survive if their group had been involved in an inter-group power struggle in the prior thirty days.[4]

What dolphins and banded mongooses demonstrate is that one way that individuals can increase their own power is via their group. Power plays out at the level of the group in feral dogs in India, capuchin monkeys in Panama, Argentine ants in California, red-tailed monkeys in Uganda, and more. And group-level power manifests itself in diverse and fascinating ways. Animal behaviorists have discovered that sometimes it is overt and bloody, other times covert and subtle. In some species, it involves males and females, in others, just one sex. It often involves control over discrete, well-delineated territories, but not always. Ethologists are busy trying to piece together the puzzle of group-level power dynamics. It's not easy.

As we delve into this puzzle, it's important to note that our discussion will steer clear of what, in evolutionary biology, is known as the levels-of-selection debate. That debate centers on whether natural selection can operate at the level of the group—favoring some types of groups over others—as well as at the level of the individual and the level of the gene. Most animal behaviorists think natural selection does not act on groups, or that if it does, the effects are minimal. For our purposes, it doesn't matter. What matters here is that intergroup interactions are important in the power landscape.

Intragroup power struggles in Argentine ants (*Linepithema humile*) make those of banded mongooses look tame. Though David Holway describes them as "a very nondescript, completely unspectacular insect," these ants live in gigantic colonies, and in Southern California, power struggles between supercolonies leave tens of thousands dead.

Supercolonies with billions, perhaps trillions, of Argentine ants can be found all over the planet, from Australia to Europe (where there is probably a single colony that spans the continent) to the United States, where they arrived in New Orleans in 1891 on a boat transporting coffee or sugarcane from Brazil. They are astonishingly good invaders, outcompeting native insect species wherever they go. Scientists studying the introduced populations observed that Argentine ants were not aggressive toward one another—just toward everything else in the path of their invading hordes—and attributed at least some of their success to that factor. Holway and his colleagues discovered that was wrong, but it took a trip to Argentina, to the native range of the ants, to figure that out.

"I was one of those kids who never grew out of his bug phase," says Holway, who grew up in the Bay Area of California. "I knew what Argentine ants were at a sort of subconscious level. They are in people's homes all over coastal California." Still, he hadn't given them any serious thought until he began his PhD at the University of Utah in the early 1990s, and even then, they initially hovered below his radar. Like many young entomologists, he had visions of his graduate school days taking him to work with some exotic species living in a lush forest somewhere in the tropics. Studies on the streets and lawns of California was "not exactly what I had in mind when I was thinking about fieldwork when I was in my twenties," Holway says. But he happened on a paper by Phil Ward on the displacement of native species by Argentine ants around Davis, California. "Invasion biology" was a new, hot field, and Holway picked up where Ward had left off, examining competition between Argentine ants and local ants, as well as the rate at which these invaders were spreading.

From Utah, Holway moved on to a postdoctoral fellowship in the lab of Ted Case at the University of California, San Diego, and a month after starting, he and Andy Suarez, a graduate student in the lab, headed down to Argentina to study the ants in their native habitat: the one place where they weren't invaders. "We had absolutely no idea of what we were doing," Holway says. "There was [almost] nothing published on the Argentine ant in Argentina," despite the fact that scientists had been studying these ants since shortly after they had landed in New Orleans 130 years ago.

Holway and Suarez soon discovered that colonies in Argentina were much smaller—in terms of the area that they occupied—than in California or anywhere else. But it was what happened between colonies that surprised them most. Everything in the literature suggested that when these ants were invaders, they displayed no aggression among themselves, but in Argentina, fights between colonies were the rule, rather than the exception.

These between-colony battles helped Holway and Suarez, and their soon-to-be collaborator Neil Tsutsui, explain why Argentine ants steamroll over native species wherever they land. Their argument goes like this: for long stretches of evolutionary time in what we now call Argentina, these ants have lived, among other places, in floodplains, like the ones at the confluence of the Río Paraná de las Palmas and the Río Uruguay. When the flooding hits, the Argentine ants, along with all the other ant species, head for high ground or float away on rafts of debris. When the floods recede, all those ant species have to reestablish their territories, and natural selection has favored traits in the Argentine ants that make them fierce competitors with Argentine ants in other colonies and with other species. When they invade new areas, they carry this evolutionary advantage, with devastating results for the native ants.

What the team's trip to Argentina did not explain was why power struggles between colonies of ants were part and parcel of life in Argentina, but no one had found such aggression among the Argentine ants in California. "If you dig up colony fragments in my backyard [in San Diego]," Holway says, "and then you go up to Neil Tsutsui's backyard in Berkeley, and put all those ants together, they would fuse and interact as if they were a single entity." Indeed, on various long road trips around California, Holway, Suarez, and Tsutsui, armed with Argentine ants from San Diego in a canister, would take samples from other populations, then place the ants captured at those sites in the canister with the San Diego ants. The ants always treated one another as if they were members of the same colony or supercolony, exhibiting virtually no aggression. On one of their road trips, with stops that included Los Angeles, Santa Barbara, San Luis Obispo, and San Francisco, they found that what has been dubbed "the large supercolony" (LSC) stretches north from San Diego for at least 600 miles. Ants anywhere on that stretch treat one another as if they were all one big happy colony.

It turns out, though, that the LSC is not the only supercolony in California. There are four others: the Lake Hodges, Lake Skinner, Cottonwood, and Sweetwater supercolonies. Each of these is smaller than the LSC, but still gigantic. The Lake Hodges, Cottonwood, and Sweetwater supercolonies share a border with the LSC, and it is at these borders that Holway and his colleagues have seen between-group power struggles on a massive, almost horrific, scale, compared with what happens in Argentina.

Holway and Tsutsui brought the *Radiolab* podcast team to one of these boundaries, between the LSC and the Lake Hodges supercolony—a boundary that happened to sit at the bottom of the driveway of a house at Eucalyptus Avenue in Escondido, California. "You didn't even need to get out of the car to see where the contact zone was," Holway says. "There were piles of dead ants along the curb." What everyone saw was 100,000 dead Argentine ant workers, the result of six months of skirmishes there. Mark Moffett, an entomologist and science writer who was along for that ride, has seen some of these battles: "They just run forward and get killed. It is quite something," he says. "The ants move forward and just start grabbing onto each other. They don't have weapons. . . . They do what most ants do, they use the rack technique and just start pulling from different directions."[5]

Not all Argentine ant boundary disputes are that gory. Holway, Suarez, and Tsutsui, working with postdoctoral fellow Melissa Smith, undertook a much more systematic study of Argentine ant between-supercolony power struggles. They mapped out boundaries, collected behavioral data, and grabbed ants to bring back to the lab for follow-up work. It was productive, but not especially glamorous, work: "We were working in the streets or the front yards," Holway says. "People would come out and ask us what we were doing."

Most pairs of supercolonies had multiple contact zones. Holway and his team stationed themselves at sixteen contact zones between the LSC and the Lake Hodges, the Lake Cottonwood, or the Sweetwater supercolony. "We removed the dead workers from the field at various times," says Holway. "There was significant mortality. . . . It was not the case that you got a pronounced battle and then a calm, [it is] more of an ongoing skirmish." When the team brought ants into the lab and staged five-against-five battles between members of different supercolonies, they found that while aggression occurred in all these battles royale, it was most intense when the ants were from areas very close to the border of their supercolonies and tapered off somewhat when the ants were from areas farther from the boundary line.[6]

Given the dangerous nature of these power disputes at boundaries, Holway and his colleagues wanted to know how the ants in one supercolony distinguished members of their own group from members of other supercolonies. That sort of question can be addressed at any number of levels. At the genetic level, while the supercolonies in California are huge today, they were probably each initiated by a small number of ants who arrived by boat, train, or perhaps car. What that means is that the initial stock of genetic variation was low, and that over time, genetic relatedness within colonies should have become high. Molecular genetic analysis done with the proper controls shows this to be the case, so power amassed at the group level tends to benefit genetic relatives. But this conclusion begs the question of what cues the ants are using to gauge kinship: how do they "know" who is kin and who is not? Part of the answer appears to be that genetic differences lead to each supercolony having a unique chemical scent: what Moffett calls a national emblem. If you secrete that scent, which in Argentine ants has to do with the production of a cocktail of what are known as cuticular hydrocarbons, you are part of the supercolony; and if you don't, you aren't.

In ants, chemical signatures are detected primarily by touch, which explains why there was a lot of touching and feeling, as well as ripping opponents to pieces, on that driveway on Eucalyptus Avenue.[7]

In many systems, power at the level of the group unfolds in less fatal, more nuanced ways than all-out violence. When it does, the underlying factors spurring contests between groups can be the most unexpected of things. For the Florida scrub jay (*Aphelocoma coerulescens*), those factors include acorns and fire.

"They need a big territory," says John Fitzpatrick, who has been studying Florida scrub jays for almost fifty years. The birds, with their pale blue plumage and underside draped in white, weigh about 3 ounces and are about 10 inches from beak to tail, but their territories, Fitzpatrick notes, are "much bigger than [those of] comparably sized birds." Part of the reason is that during the winter, scrub jays survive almost completely on the eight thousand or so acorns they bury over the rest of the year. That alone requires lots of storage space, and to make matters more difficult, fire periodically sweeps through and burns their low-growing scrub oak habitat, so that the trees don't produce acorns for two or three years. That means scrub jays need a territory large enough that some of its trees are not hit by flames. "Territory is everything," Fitzpatrick says. "The basic formula is to defend all the habitat that you can possibly defend. The problem is that everyone else is doing that, too." And that's what leads to power struggles between groups.[8]

In 1972 Fitzpatrick was an undergraduate at Harvard, planning out the upcoming summer. "Right at the moment when I was thinking I don't want to mow lawns again, I want to do something that has to do with science," Fitzpatrick says, his roommate told him of an advertisement he had seen on campus for a summer internship at the Archbold Biological Station in Florida. He applied and was accepted to work under Glen Woolfenden, who a year earlier had begun marking Florida scrub jays at Archbold. Fitzpatrick's work that summer looked at power and dominance relationships *within* families of scrub jays. He built "a bunch of little gizmos that allowed only one bird to be feeding at a time," and then could see who won access to peanuts in a feeder. When Fitzpatrick went down to Archbold again the next summer, he became more and more interested in power and territoriality among the scrub jays. He started mapping out territories and found that family groups, which included breeders and helpers, were defending territories year-round, and that the boundaries between the territories were very precise, on the order of a yard wide.

One of his roles from very early on was what he calls "field marshal for the annual territory mapping process." As Fitzpatrick and his team mapped out territories, he came to learn that he could create between-group contests to better study them. The birds knew him and his assistants well, because, he says, laughing, "we were an occasional source of peanut morsels." He'd go into the middle of one territory and mimic a scrub jay vocalization: "The local scrub jay family comes over, in the hope of getting a peanut," Fitzpatrick says, ". . . and we move them over to the edge of a territory, and we do the same thing to a neighboring group. . . . They come together . . . they forget about food. They fight." What he and his team found was that territorial borders were not only very precise, but could run 200 meters long.

Members of families on adjacent sides along that 200-meter boundary defend their territories and, if things go their way, encroach onto their neighbors' land, and perhaps even take over a little chunk of it and make it their own. In these battles, scrub jays use an array of aggressive behaviors, including vocalizations, threats, chases, and, on occasion, more serious grappling, where birds grab each other's legs and roll on the ground, until the winner began pecking the loser. Male breeders were more likely than female breeders to take part in these skirmishes, and male helpers were far more likely than female helpers to join in.

Because territory size is the key to success in the acorn-limited, fire-prone world of the Florida scrub jay, these between-group struggles for power really matter. Winning—and even more importantly, usurping a bit of the neighbors' land—can make the difference between having enough acorns to survive the winter or not. Fitzpatrick and his team found that larger groups tended to defeat smaller groups, and slowly they pieced together how family groups were augmenting their troops and increasing the probability of winning power struggles along the border.

Male scrub jay helpers tend to provide more assistance to breeders than do female helpers, and they remain on their natal territory for many years, while female helpers disperse at a relatively young age. The payoff to male helpers for staying can be sizable. First, by helping, they increase family size, which increases the probability that they and their kin will hold the family land. But there's another benefit as well. In rare instances, after a male breeder dies and his mate disperses, the dominant male helper inherits the entire territory. More often, the dominant male helper will settle on the periphery of the family territory and push the boundaries out slightly, carving out a small realm of power for himself and a female he is courting: "Males bud off a piece of Mom and Dad's territory," notes Fitzpatrick, "and inherit the back forty, so to speak."

Budding is something Fitzpatrick and his team constantly keep an eye out for. Every year, interns coming running back from the field to tell him of some observation suggesting that a budding event may be unfolding. That gets everyone's attention. "We light up," Fitzpatrick says. "We go back there several days and watch the progress. . . . When it works, a male has carved out a five- to eight-acre piece of scrub; when it doesn't work, [it's because] neighboring families are constantly fighting [with him] and [he] gives up."

Half the time, budding is successful, and half the time it's not. When budding works, it benefits the helper, for obvious reasons; but his family benefits as well, because family territory size increases. That translates into both more acorns and a larger group size — particularly when the pair on the budded area produces offspring — all of which augment group size for future power contests with the families next door.[9]

For the street dogs (*Canis lupus familiaris*) of Kolkata, it's not acorns and fire, but trash dumps and food vendors that largely drive contests for power between packs. No one knows that better than Anindita Bhadra, of the Indian Institute of Science Education and Research, who leads a team studying their social behavior.

"I was always interested in these dogs," say Bhadra, who grew up in Kolkata. "As a kid I used to go feed street dogs outside and go fight with boys who threw stones at the dogs." But years later, when she entered a PhD program in animal behavior under India's leading ethologist, Raghavendra Gadagkar, Bhadra chose to work on the social system of *Ropalidia marginata*, a wasp species that Gadagkar had been studying for decades. Her dissertation, "Queens and Their Successors: The Story of Power in a Primitively Eusocial Wasp," focused on queen succession. In particular, queens in this species show much less aggression than those of other wasp species, yet maintain complete control over reproduction. When the queen dies, succession is rapid: another female quickly becomes very aggressive herself and assumes the role of queen. But once she does, she, too, becomes docile and nonaggressive. In contrast to many other social insect species, which female becomes successor is not a function of dominance rank within the hive, age, or size. Bhadra's dissertation pieced together the cryptic rules of how succession unfolds.[10] "I used to say, 'I study wasp politics,'" quips Bhadra.

When she finished her dissertation, Bhadra had to make a difficult decision. She enjoyed working with the wasps and could have kept on working in that system, but in the end told Gadagkar, "Nobody can do that better than you." Right at that time, a slew of assistant professor jobs had opened at the new Indian Institute of Science Education and Research (IISR) in Kolkata, and they were hiring fresh PhDs. Bhadra wrote up two research plans for the position. One centered on the social behavior of crows. The other brought her back to the street dogs she knew so well from her childhood.

"I was reading a lot of papers on dogs," Bhadra recalls, and "it struck me that people were talking about the evolution of dogs, social cognition, and human-dog interaction, but they did all of these experiments on pets. I was a bit upset about this." That was not what life was like for street dogs in Kolkata, or anywhere in the world. "[Pets] are constantly under human supervision," Bhadra continues. "It was different [in] the dogs I had seen fighting for every scrap . . . and I thought, 'Why doesn't someone study free-ranging dogs in what we think of as a natural habitat, because they have lived like this [in India] for centuries.'"

She went to Gadagkar for advice with both the crow and dog plans in hand. "He said 'both are good,'" Bhadra remembers, "'but where is your heart?'" It was with street dogs. What's more, from a practical perspective, a job at IISR would come with heavy teaching responsibilities. The street dogs were everywhere, the crows weren't, and they were much harder to work with than the dogs.

About eight hundred million of the billion or so dogs on the planet don't live in homes with the species that domesticated them. The dogs that Bhadra and her team work with around Kolkata have not escaped a human household or been heartlessly discarded from one. They are typically called free-ranging or free-living dogs, in the sense that they don't live with people, nor have their ancestors for hundreds of years. Still, they rely almost entirely on humanity writ large for food, through scavenging at trash dumps, grabbing scraps around food vendors, or being fed by locals—many an Indian folktale tells of their escapades. The animals that Bhadra studies are also called street dogs, but some don't live on streets proper (though the ones in Kolkata largely do).[11]

Bhadra and her team of undergraduate and graduate students study everything from feeding behavior and mating strategies to parental care, aggression, and power dynamics. In Kolkata, a pack's territory often maps nicely onto one side of a particular street, with another territory often on the other side. Bhadra's team knows all their dogs by sight, and they use a combination of pen and paper and video camera to gather data. Because of the pounding daytime sun, they tend to do their data-taking early in the morning and in the late afternoon. The heat gets to the dogs as well as the scientists: "Most of the time the dogs are just lounging around, they are not doing anything," says Bhadra. "The interaction rates are low, but bonds [between pack mates in a territory] are very strong. . . . They are watching each other. . . . If A does something like barking, B mirrors it. . . . We are seeing very subtle leadership cues."[12]

Urban fieldwork like this, especially with creatures like dogs, has its own challenges. Many of Bhadra's assistants are undergraduates. Almost all of them are dog lovers, and many have had no coursework in animal behavior. "I have to go through a whole process of telling them that if you are doing observations, you cannot pet the dog,"

Bhadra says. "You cannot cuddle it, you cannot feed it. You cannot rescue it, even if it is dying . . . you cannot bias your data."

Bhadra and her team do their best to prevent the dogs from becoming too acclimated to them. But they are street dogs, and they interact with people all the time. "We get a lot of interference when we do our experiments," Bhadra notes. "People will ask, 'What are you doing to our dogs, what are you feeding our dogs, are you poisoning our dogs?'" Some, the dog lovers on the street, want to make sure "their" dogs are not being mistreated; others, who see the dogs as nuisances and vectors of disease (for the most part, they aren't), would be happy to hear that Bhadra is putting out poison to rid the streets of them. Most people lie in the middle and don't especially care about the street dogs one way or the other.

Street dog social systems vary from place to place around the world, but among the Kolkata street dogs that Bhadra studies, there is no obvious dominance hierarchy within packs. "I think it is kind of beautiful," she says with a smile, "a kind of democratic system." But it is different when it comes to the power dynamics between groups. The street dogs are not nearly so laid-back when it comes to their territories. They aren't mongoose or Argentine ant nasty, but they don't take well to attempts at land grabs.

Pack territorial boundaries are clearly marked with urine, and each night borders are reinforced by prolonged and vociferous bouts of barking that can drive the humans living on that street crazy. Usually dogs in packs on different sides of the street respect the boundary dividing them. But sometimes there is an incursion by one or a few dogs from one side to the other. A series of chases ensues. If the intruders remain, fights sometimes follow. In the vast majority of cases, it ends there, with the intruders leaving and the border unchanged. The interaction has an eerie feel to a human watching it all, as Bhadra describes it: "There will be groups lined up on different sides of a T-junction on a street, and they are just standing and there is a match going on and they are just barking and suddenly they decide it is over and they just go over and lie down."

As ethologists have seen in so many social systems, the more valuable a territory, the more likely that aggression over it will escalate, sometimes very quickly. And because dogs are about as cognitively complex an animal as there is, sometimes researchers just can't quite figure out the dynamics of intergroup power struggles, even when aggression is ramped up and obvious to all the dogs involved and all the humans watching. As a case in point, Bhadra tells of an open field where two packs, one large and one smaller, have territories on each end. It's valuable land to the dogs because there are food stalls nearby that provide lots of scraps. "The two groups used to fight constantly," Bhadra says. "Whenever the males met, they always had vicious fights. Two dogs died in these fights."

Bhadra and her team are almost always able to assign a dog to one pack, and only one pack, but on this open field, they were stumped by one black-furred male, who came and went between packs as he pleased, tolerated by all. He mated with females in both groups and could move between groups and play with the pups at will. In principle, this male could have served as a liaison of sorts, leading to reduced violence between these packs, but that's not what happened. "When the two groups met and fought, everybody, males and females, would be fighting," Bhadra says, "and this black male would sit on one edge of the field quietly and watch like he was watching a tennis match. Never participated in a single fight. We used to call him Gandhi."

Why Gandhi did what he did, and why the others in both packs allowed him to do it, is unclear. But he and his packs have prompted Bhadra and her team to try to figure out new ways to probe more deeply into the subtle dynamics of between-group power struggles in the street dogs of Kolkata.

Tales of between-group power struggles in primates are legion. But it's not easy to study such struggles, let alone set up experiments. When you're a primatologist, Meg Crofoot says, "you work very hard to habituate a single community; you don't usually have the luxury in most cases of studying the behavior between groups. . . . It is limitation of manpower and time, not interest." But as a PhD student in the early 2000s, Crofoot had the interest and the time. She was planning on studying power (and lots of other things) in white-faced capuchins (*Cebus imitator*) on Barro Colorado Island (BCI) in Panama.

Crofoot shared an apartment with fellow graduate students at MIT. One night, when they were all at a pub, Crofoot was thinking out loud and said she'd need an army of assistants to help her track the groups of white-faced capuchins she was hoping to work with at BCI. Her techie apartment mates thought otherwise and told her what she really needed was an automated tracking system. But the more she looked into it, the more Crofoot realized that building such a system would be an entire dissertation project in itself, never mind using it for a study on capuchin behavior. Serendipity struck when she discovered that Martin Wikelski and Roland Kays, who Crofoot had met a few years earlier, were building exactly what she needed—an automated radiotelemetry system they dubbed ARTS—and that ARTS was to be shared with any and all researchers on BCI.

Groups of capuchins on Barro Colorado Island range in size from nine to twenty-five individuals. To start her work, Crofoot asked her colleague Bob Lesnow to dart one or two monkeys from each group with an anesthetic called Telazol and place radio collars on them. That process went smoothly, though the capuchins, reasonably enough, did not like it one bit. One monkey, who Crofoot dubbed Bravo Louis, would have no part of it. Two days after they collared him, he broke the antenna off his collar. "I have no data on him. But he is alive," says Crofoot, who just wants to get the collar off poor Bravo Louis. "It is now seventeen years later. . . . It's amazing, he sees me and it is all good. He sees Bob [who darted him]—I mean all Bob has to do is set foot on the island—and Bravo is up in the tree and across the island immediately."

ARTS was soon sending Crofoot location data on collared individuals every ten minutes, twenty-four hours a day, and because capuchin group structure is very cohesive, that told her where their groups were as well. But it didn't tell her anything at all about what a particular group member was doing. For that, Crofoot and her assistants used more traditional focal individual sampling, rotating between groups for three-hour stints to note who was feeding, grooming, displaying aggression, and so on. Those were long days, often starting around 4:30 a.m. She'd start with a quick breakfast and a peek at the ARTS data to find out where the groups were that morning. "Capuchins are exhausting animals to watch," Crofoot says with a smile. "They are always doing things." Part of what they spend a lot of time doing is shredding everything looking for fruit and insects. "I think that their manipulative foraging niche translates into manipulating everything else," she continues, "so their interactions with other species are frequently quite funny. Picking up baby coatis and twirling them by their tails like a lasso and sending them flying. Capuchins have never seen another animal in the forest that [they haven't] harassed. Including humans."

Neighboring capuchin groups have fairly well-defined boundaries, but there is usually a 20% zone of overlap. And that's where most (though, as we shall see, not all) intergroup interactions occur, including, on average, an intergroup conflict every three days. "Sometimes a group simply turns around and vanishes as soon as they realize another group is nearby," Crofoot says. At other times, capuchins from both groups rush at one another up in the trees, with "the big males climbing up high and bouncing on big branches until they break and come crashing to the ground, males and females doing coalitionary displays, where they climb on each other's backs and stack their heads like a totem pole while making threat faces at members of the other group." If that fails to make one group leave, individuals from each group "line up on the ground like American football teams, lunging and screaming at each other and chasing each other," Crofoot continues. "In the most intense cases, you often have even females with small infants clinging to their backs right in the middle of it." Knock-down, drag-out fights are rare, but they do happen.

With a general feel for what between-group encounters look like, Crofoot began to piece together capuchin power dynamics by using the ARTS location data. That might sound like an indirect way to assess who wins intergroup encounters, but it turns out that it's a good proxy for direct observation. "When two groups encounter one another," Crofoot explains, "one leaves [loses] and the other stays [wins], or they both leave. There are no ambiguous outcomes."

Which means the ARTS data work just fine for studying some things at the level of the group.[13]

One thing the ARTS data showed is that there are costs to losing intergroup power struggles. Individuals in losing groups spent more time traveling around looking for the fruits, nuts, and insects that make up the majority of their diet, and they were more likely to end up foraging in low-quality food patches. In addition, individuals in winning and losing groups showed differences in the number of sleeping sites they used.[14]

Crofoot also used the ARTS data to get a handle on what determines the outcome of a contest between two groups and, in particular, how group size and location affect these outcomes. Using data from fifty-eight intergroup encounters in which one group clearly displaced another, Crofoot and her colleagues found an interesting interplay between group size and location. "Big groups are more likely to win than small groups," she says, "but you have this interesting thing where . . . for some reason, when [big groups] invade the territories of their small neighbors, they [the big groups] are not winning there." More specifically, what she found was that every additional group member increased the chances of winning a group contest by 10%, but that small groups near the center of their own territory were still very much a force: for every hundred meters that a large (or, indeed, any) group moved from the center of its home territory, its chances of winning an interaction with another group decreased by almost a third.

One reason why small groups might win when they are deep in their home territory is that the land is worth more to them than it is to the invaders. It takes time and effort to learn about the territory you are living on—where the food is, where the predators aren't, and so on—and that means that the same area of land is worth more to an animal who has invested in learning the lay of that land than to one who has not. But Crofoot wondered if there was another reason why small groups defeat large groups who invade their territory. Could it be that large groups weren't really as large in practice as a head count indicated, because not everyone in those groups was doing their fair share? The ARTS data, combined with Crofoot's data on group size, useful as they were, couldn't address that possibility. But there was theory that she could turn to.

Crofoot knew the game theory literature on collective action and what is called the free-rider problem. Extensive modeling by evolutionary game theorists had long since confirmed what economic game theorists had found earlier: in groups, natural selection always favors individual group members who reap the benefits of cooperation, but sidestep paying any of the costs associated with cooperation. If some pay the cost of cooperation, and others can procure the same resource but cost-free, then "free-riding" cheaters can thrive. Perhaps, Crofoot thought, free riders were another reason why large groups fared poorly against smaller groups defending the core of their territory. The best way to know was to do field experiments in which free riding, if it occurred, could be directly measured. Playback experiments, she decided, were perfect for that.[15]

Crofoot and her colleagues created one-minute-long audio files from recordings of four of their six study groups. The audios had sounds associated with foraging, including food calls, falling fruits, and monkeys on the move. Halfway through each audio, screams associated with fighting were inserted. Sounds from all group members were included, providing a rough estimate of group size to listeners. Next, capuchins in a given group were played audios from another group to simulate a territorial intrusion. In some cases, the audios were broadcast from a speaker placed in the center of the listening group's territory and, in others, from a speaker placed at edge of its territory. In all cases, the speaker faced the listeners from the direction of the group whose sounds were being broadcast.

A capuchin was categorized as approaching, and hence likely to be ready to participate in, an intergroup encounter, if in response to the playback, it left the tree it was in and moved 5 meters in the direction of the speaker. If a monkey left the tree and moved 5 meters in the opposite direction from the speaker, it was categorized as retreating from a possible intergroup power contest.

Capuchins were almost seven times as likely to approach a speaker if it was placed in the center of their territory than if it was placed at the edge. Free riding was strongly tied to location: the odds that a capuchin would retreat—that it would free ride on others—were 91% greater at the edge of its territory. So, large groups venturing from the heart of their territory were especially susceptible to free riding, which, as Crofoot had hypothesized, explains in part why it was that small groups defeated much larger groups when territorial incursions were deep and much was at stake.

At the Ngogo Research Station in Kibale National Park, Uganda, Crofoot's colleague and friend Michelle Brown has been studying a primate system in which power at the group level plays out in different, but equally intriguing, ways.

For the last fifteen years, Brown has been studying red-tailed monkeys (*Cercopithecus ascanius*) in remote areas of Kibale, away from the tourists who are now all too common at the heart of the park. She works with a large group of Ugandan field assistants that she spent months training, and they follow red-tailed monkeys (as well as other primate species) with a particular eye toward mapping out intergroup power dynamics.

Red-tailed monkey groups defend home ranges, but, as with the capuchins, there is often a zone of overlap between ranges. Power struggles usually happen in those zones, most often over a good patch of food, though "why this fig tree and not that fig tree," Brown says, "I don't know." At certain times of the year, neighboring groups average one encounter a day, and Brown quickly got quite good at predicting when a fight was looming. "I like to say that I speak monkey," she says, "and so I know when they are going to have one of these fights." When she thought that was about to happen, Brown quickly stationed assistants at the forward and rear edges of the overlap zone, and around where the fight was about to unfold, and they "would watch what was going on, communicating with each other, while we were spread out."

Though she has not seen monkeys seriously injured during intergroup encounters, these power contests do get nasty: "If they can grab an individual of the neighboring group," Brown says, "they will slap it, bite it, and sometimes knock it out of a tree." It seems reasonable to assume that these sorts of encounters would be stressful, but almost no one studying group-level power had tested whether that was in fact the case. So Brown used the hormones floating around in red-tailed monkey urine to do just that.

Brown and her colleagues collected urine samples by pipetting them up from vegetation after the monkeys had deposited them there. They stored the samples in a solar-powered freezer until they could be shipped overseas for analysis of, among other things, cortisol, a hormone associated with stress responses. Of those urine samples, one-third were collected during or shortly after an intergroup contest. The other samples were used to gauge baseline levels of cortisol. In total, the team collected 108 samples from twenty-three red-tailed monkeys, and for each of these monkeys except one, they had both baseline samples and samples from an intergroup encounter.

What Brown found was elevated levels of cortisol in samples from intergroup contests, which is just what she expected to see if contests were stressful. But there were two unpredicted twists to the story. For one, even though many studies on aggression have shown that losing a fight increases cortisol levels more than winning a fight does, the increase in cortisol levels in the red-tailed monkeys was the same regardless of whether an individual was in the winning group or the losing group. The other twist was that the increase in cortisol was significant only in the samples collected during, or shortly after, an intergroup encounter, not in the samples collected a few hours after an encounter. That surprised Brown, because work in other species of primates had found that the "the prime excretion window" for cortisol was a few hours after some stressful event (like aggression), not during or soon after the event. And yet she was seeing the spike in cortisol at the latter time, but not the former. What, she wondered, could explain that? It might be that the prime excretion window for cortisol in red-tailed monkeys is much closer to the stressful event than in other species, and that the monkeys were indeed stressed by the power struggles per se. Another possibility is that the monkeys had detected signs that an intergroup encounter was imminent, perhaps long before Brown and her team had, and the spike in cortisol was an anticipatory response to an upcoming stressful contest for power. Brown's urine samples do not allow her to distinguish between these hypotheses, though both are equally intriguing when it comes to understanding power at the level of the group.[16]

Power struggles between groups, power struggles between individuals in groups—everywhere, power matters. But even when animals do everything they can to attain power and then cement their position, power is a precarious commodity. Sometimes power structures crumble, only to be rebuilt in new ways.

8   Rise and Fall

No one ever seizes power with the intention of relinquishing it.
GEORGE ORWELL, *Nineteen Eighty-Four*

Animals are always looking for the chance to force others to relinquish their power and make it their own. Ravens in the Austrian Alps certainly are.

Raven politics appears to involve more than intervention and the audience effects that we saw in chapter 4. Thomas Bugnyar, Jorg Massen, and their colleagues learned that ravens can also detect when the balance of power between *others* is shifting. Bugnyar and his team exposed a raven to a recording of two other birds from its group. In some trials, that recording had the sounds of a higher-ranking bird making vocalizations associated with dominance ("self-aggrandizing displays," or SADs) and a lower-ranking bird emitting sounds associated with submissive behavior. From the point of view of the bird hearing these calls, no expectations about power relations among others were violated: dominants were making calls dominants normally make, and subordinates were making the vocalizations typically associated with submissive behavior. But in other trials, Bugnyar set it up so that the lower-ranking bird was heard emitting SAD calls and the higher-ranking bird making submissive calls, suggesting a possible shift in power. When that happened, female ravens increased "self-directed behaviors" associated with reducing stress levels, hinting that shifts in the power structure are anxiety-provoking.[1]

Ravens, like so many other animals, are continually collecting information and updating the costs and benefits at play in their power structure. This updating is important, because sometimes, for any number of reasons, including nosy researchers tinkering with the natural order of things, power structures may dissolve and then form again among the same set of individuals.

When Allen Moore and I asked Michael Alfieri if he would be interested in joining a project we were starting up on the dynamics of power, he didn't think it would involve working alone in the *Twilight Zone*–like setting of a small, cramped, dank, dark room swarming with cockroaches. He was very wrong.

It was 1992, and I was a postdoc in the evolutionary ecology group at the University of Kentucky. Moore was a professor in the School of Agriculture on the other side of campus, and Alfieri was just starting his PhD work. Moore had done work on dominance hierarchies in speckled roaches (*Nauphoeta cinerea*), and I had a burgeoning interest in the evolution of power; together, we had hatched an experiment that married our interests. Alfieri and I had worked on some other projects, and I thought he would make a good addition to the team. He agreed: "Roaches . . . why not?" Alfieri recalls thinking. "Just another model system to test questions about social evolution."

When we started, there had been a grand total of one controlled experiment ever done on the replicability of dominance hierarchies—that is, whether the same power structure emerges if a hierarchy breaks up and then re-forms. That experiment had been in 1953 and involved a single group of chickens. But speckled roaches were a perfect system for remedying the lack of work on this topic. They form strict linear hierarchies, and their aggressive acts—like butting, lunging, biting, kicking, and grappling—are easy to record, as are their submissive behaviors, like walking away, crouching, withdrawing, and retreating. What's more, these roaches are easy (and cheap) to work with, and Moore had large breeding colonies on campus.[2]

Alfieri's main task was to lead a "form re-form" experiment in a small room where we had switched the day-night cycles so that he could watch the nocturnal roaches by an eerie red light that the roaches were color-blind to. First came the prep work. Alfieri vividly recalls "hours and hours of gluing tiny numbers on the backs of roaches." We made up eleven groups, each with four marked male roaches that were roughly the same age and size. Each group was in its own shoebox-sized plastic arena that had petroleum jelly rubbed along the top edges to make sure the roaches stayed put. Housed along the walls of the room were hundreds of roaches that Moore had other plans for.

As soon as a group of four roaches was placed together, Alfieri was there, observing their behavior and calling it out into a tape recorder: "Roach 1 lunged at 2," and so on. "I have vivid memories of that room," Alfieri says. "I remember the sound of hundreds of roaches scratching on their containers while we were watching four roaches interact in an arena. . . . The room was hot, and I was in shorts and T-shirt and constantly brushing off what I thought were roaches crawling on me. . . . You'd just feel them on you, although none of them ever got out. Twenty years later and I still remember that sound and that feeling."

All in all, it was an eight-day experiment. In round 1, every day for three days, Alfieri gathered data on aggressive and submissive acts in each group. In ten of the eleven groups, a clear linear hierarchy emerged. Then each of the males in those ten groups was isolated in its own home box and given food and water for two days. In round 2, the groups were re-formed, and for the next three days, Alfieri recorded the same sort of data he had earlier. When we looked at data from those ten groups, all had formed linear hierarchies in round 2. But replicability was spotty at best: five groups had re-formed the same hierarchy in round 2 as in round 1, and four of them had not. In other words, sometimes disruptions shook up a hierarchy and sometimes they didn't.

We searched for threads that linked the groups in which the same hierarchy re-formed. The only hint we got was that if we looked at the top-ranked males in round 1, they were involved in a smaller proportion of interactions in groups where the same hierarchy was re-formed than in those where it was not. Exactly why, we don't know, and in the end, we were left scratching our heads about what distinguishes groups that rearrange the power order differently on a second go. But the very fact that there were now actually some data on replicability of dominance was a step in the right direction.[3]

Power structures are often stable for long periods of time. But speckled roaches show that they are also precarious because they depend, in complex ways—some of which we don't yet understand—on a real-time updating of their costs and benefits. Sometimes the power order crumbles, only to be rebuilt in new ways, as in macaques on Cayo Santiago Island, mantled howler monkeys in the fragmented forests of Veracruz, and cichlid fish in Lake Tanganyika. Animal behaviorists working in these systems are trying to piece together what leads to revolutions from within and takeovers from without, how these events affect everything from hormones to gene expression in the animals involved, and what happens to the deposed, formerly powerful individuals, as well as those that rise up to take their place, when a new power order rises from the ashes.

For the most part, power in rhesus macaques (*Macaca mulatta*) revolves around family connections, ritualized displays, moderate aggression, and lots of patience waiting for those above you in the pecking order to die. But not always. Sometimes all hell breaks loose. "To change the power structure of a group, normal fighting is not enough," says Dario Maestripieri, who has been studying these monkeys for thirty years. "To induce an alpha female who has been in power for ten years to relinquish her position, [another macaque] almost [has] to kill her and her family members, because they will essentially fight to the death to protect their power."

Maestripieri did his postdoctoral work at the Yerkes Field Station, about 20 miles outside of Atlanta, which is home to more than a thousand rhesus and pigtailed macaques housed in large outdoor corrals. Everywhere he turned, the rhesus macaques would be positioning themselves to acquire and maintain power. There were even full-fledged revolutions, and they were not pretty.

Among rhesus macaques at Yerkes, much of the power runs through female lineages. Within a group of a hundred or so monkeys, the largest matriline is usually the most powerful, the second largest matriline is next in line, and so on. At one point, the dominant matriline at Yerkes shrank in size, and that spelled trouble. "The second-ranking matriline outnumbered [them]," as Maestripieri tells it. "Then there was essentially a revolution: just one day, out of the blue, a big war, in which the second-ranking hierarchy . . . attacked all the higher-ranking females to kill them. It was a bloodbath. . . . During these matriline overthrows, they attack to kill . . . biting them in their faces and their necks."

These revolutions often begin in a rather unspectacular way. There will be some sort of aggressive interaction between members of the second-ranked and top-ranked matrilines: nothing especially violent, just the normal sorts of aggressive acts—threats, charges, chases, and so on—and submissive behaviors—cowering, bare-teeth grinning, avoidance—that happen all the time. The combatants' kin quickly take sides, which again is par for the course. Soon the top-ranking female in the top-ranked matriline joins the fray. Usually, things settle down quickly at that point. But when the top-ranked matriline has shrunk in size, monkeys from the second-ranked group don't back down. "They manage to recruit more and more allies," Maestripieri says, "and in the end they win. . . . The outcome is that there is a revolution. It typically starts in the evening when things are quiet, and when we get there in the morning, there are dead bodies on the ground."

When the top-ranked matriline loses its hold on power, its members don't just fall one notch in the group hierarchy, they plummet. "Females who become outranked drop to the bottom of the hierarchy," says Maestripieri. "The alpha female becomes the lowest-ranking female in the group—if she survives . . . all her relatives also drop to the bottom."

Maestripieri came to learn that these sorts of revolutions occur, in a slightly muted form, in males as well. After he left Yerkes in 1999 to take a faculty position at the University of Chicago, his friend and colleague Melissa Gerald suggested he move his rhesus research to Cayo Santiago, a small island off the coast of southeast Puerto Rico. Cayo Santiago is home to anywhere from five to ten groups of rhesus macaques. The monkeys all descend from animals brought from India in 1938, and they can roam freely on the island most of the year. Researchers like Maestripieri and his students, and usually a half dozen other scientists, take the ten-minute ferry ride each morning from Punta Santiago, a small town where they rent housing, to the island. At the end of the day, all humans leave the island.

Armed with binoculars, pen and paper, and occasionally a video camera, Maestripieri, Alexander Georgiev, James Higham, and their colleagues witnessed two power shake-ups: one from within and one from outside. Males on the island sometimes leave their group and migrate to another. "Traditionally, when a male enters [a group], they enter at the bottom of the hierarchy," Maestripieri says. "They are very submissive and over months or years, they slowly work their way up the hierarchy." Male 11Z from Group R wanted no part of that tradition. This nine-year-old male had been the alpha male in Group R when he moved to a neighboring group, Group S, in early March 2013. By March 15, he was the clear alpha individual in Group S, winning 96% of contests in which he challenged, or was challenged by, one of the other males in the group. Even when coalitions of males came after him, 11Z didn't back away: "On [our] video you can see [11Z] being surrounded by and mobbed and attacked by the resident males and he just won't stand down. He stood his ground and he fought and eventually he was able to defeat this gang of males."

Male 11Z mated with more females than any other male in his new group, but then something happened. Other males in the group were still fighting 11Z's takeover, and by July 17 he had been deposed as the alpha male and crashed to the position of lowest-ranking male in the group. Though Maestripieri and his team did not see the events leading to his downfall, it clearly involved severe fighting, as 11Z now had nasty gashes on the back of his head and on his inner thigh, and a deep wound just above his testicles. From that point on, other males, including—indeed, primarily—those very low in the power structure persistently targeted him for aggression. As the lowest-ranking male, he had low priority at feeding sites and lost nearly 15% of his body weight in a matter of months, more than any other male in the group. What's more, blood samples showed that 11Z had the highest levels of neopterin, an indicator of inflammation and infection, of any male in Group S.[4]

Other times, male rhesus macaque coups are homegrown and involve what Higham and Maestripieri call "revolutionary coalitions." In one group they were tracking, they observed numerous coalitions formed by middle-ranking group members who targeted males in their group who were above them in the power structure, including alpha and beta males. These coalitions could be relentless: in their paper "Revolutionary Coalitions in Male Rhesus Macaques," Higham and Maestripieri described what happened to male 83L, the beta male in his group: "First observations, 22 June, involving [males] 57D, 44H, 50B. [Male] 83L was driven from the group into the sea. Sustained attacks over the following two weeks, in which 83L was chased all over . . . and repeatedly driven into the sea. . . . Coalition members sometimes changed between observations. [83L] was eventually allowed to return to group periphery in a lower-ranked position."

Within just a few months, all the males targeted by coalitions had dropped rather dramatically in rank, and the (former) alpha male was forced to leave the group permanently. Almost all members of coalitions rose in rank during the corresponding period.

Power has its rewards, but in rhesus macaques, it targets you for trouble from upstarts.[5]

In Maestripieri's macaques, solo takeovers come from the outside while coalitionary takeovers come from within. There are other permutations, though—other ways to shake up the primate power order. In the mantled howler monkeys (*Alouatta palliate*) that Pedro Dias studies in the forests of Los Tuxtlas, in Veracruz, Mexico, that can come about when a coalition migrates in unison and shares the reins of power in its new domain.

"When I was a kid in Portugal, we only had one TV channel and

Saturday morning . . . [was the only time] we had documentaries on animals," says Dias. "I always wanted to work with animals. The thing was I had lots of other interests and I was a really bad student." So bad that when it came time to go to university, he had no chance to get into a good biology program. But, as Dias tells it, at the time it was relatively easy to get into an anthropology program, and since going to university was expected of him, and since some anthropologists study nonhumans, he took that route. At first, he was bored, "but then I [took] this amazing course on biological anthropology. . . . It was a revelation for me," he says. "I could get a degree in anthropology but then go and work with animals."

In addition to healthy doses of evolution, population genetics, and animal behavior, that course had a field component. At first Dias did a project at Lisbon Zoo, but then a master's student working with mantled howler monkeys in Mexico came to give a guest lecture in the class. Dias was hooked. He did a master's project himself on the monkeys on a small island in Veracruz, but that population was a translocated one with inbreeding issues, so for his PhD work he went to Los Tuxtlas, where a long-term study of mantled howlers was already underway.[6]

The area of Los Tuxtlas where Dias works is a fragmented forest region on private property. For the monkeys, that means "movement through landscape is strongly influenced by humans," says Dias. "The monkeys have to figure out where to move through landscape, because there are lots of pastures. . . . To get to some fragments they have to go to ground and walk." And they don't like leaving the trees to do that. The landscape also presented its own special hurdles for an aspiring PhD student. "Sometimes people decide it is OK for you to work with monkeys [in their part of the forest]," says Dias, "and then one day they decide you have to pay them something to be there." But the mantled howlers were everywhere, and the setup was nice, so he rented a house with another PhD student in the small town of Montepío.

After a 4:30 a.m. breakfast, Dias, led by the beam of his flashlight, would walk through the dark forest each morning to the last spot where he had seen a group of monkeys the prior evening. The monkeys move from tree to tree during the morning and afternoon, feeding as they go, and they usually sleep in the last tree in which they feed, so there was a good chance Dias would find them. But "if you miss that first movement, then it may be very hard to find them," he says, and "that will be a day you won't collect data."

Dias watched groups of mantled howlers in the trees through his binoculars as he spoke into a tape recorder. The animals were all recognizable by fur pattern and body markings. Proximity data—where the animals were in relation to one another—was tricky, but important, to record, because the absolute number of times howlers interact with each other is low for a primate, so where they are when they do is critical information. His solution was simple, but elegant: "I knew the length of the arm of an average monkey was about 35–40 centimeters, so I started extrapolating from that."

For the most part, aggression within mantled howler groups is fairly muted. By far the most common aggressive interaction is a supplant, in which one monkey approaches another, and the other howler simply leaves. Sometimes there is a bit of shoving. Other times, Dias says, monkeys will "arch their back and shake the branches. . . . They break branches and they break them again and they vocalize while they do this." On rare occasions, there will be a real fight that might involve some biting, but almost never any serious injury.

Between groups, things are also fairly calm in Dias's mantled howler population. "They are all about avoiding conflict," he says with a smile, ". . . they have these calls that allow them to track neighboring groups so they [can] avoid them." If groups did end up interacting, usually it was innocuous enough: "It [can be] really funny, because you have the males from two groups howling and shaking branches and doing all their displays. And then you have all the little ones [the juveniles] coming, playing in between them."

Other times, it was not so funny. Dias recalls one male who left his group and was gone for three years before returning with a mate. He decided to set up his home range in the middle of the home range of one of the groups already there. "He really beat up one of the resident males. He threw him to the ground from the trees," Dias says.

"These moments that are associated with takeovers [can be] really violent."

In December 2003, Dias witnessed a takeover unfolding before his eyes when a pair of males invaded the MT Group of howlers—made up of two males, four females, and five younger individuals—that he could hear from the porch of his house in Montepío. The normal calm of mantled howler life fell away as the pair beat one of the residents to death. Nearby was another group of howlers (RH) that was similar in composition to the MT Group, but had not been invaded by outsiders as MT had, and thus provided Dias with a natural control for comparison purposes as he studied the aftermath of the invasion.

The general level of aggression in MT was much higher than in RH, and Dias found that serious aggression, such as fights and bites, was absent in RH, but common in MT. What was especially striking to Dias was that the pair that invaded MT acted as a well-oiled coalition. They howled together, often attacked others together, and exchanged more friendly interactions with each other than did any other members in group MT or RH, which is all to say that shake-ups in power structure sometimes require individuals to cooperate in ways they might not otherwise.

These dramatic shifts in power are not unique to primates, or even to animals dwelling on land. They happen underwater, too. And when they do, at least in one of the cichlid species that calls Lake Tanganyika home, they change everything, from behavior to hormones to genes turning on and off in the brain and testes.

"When I did my PhD out in Hawaii, I was working on an endemic damselfish species," Karen Maruska says, "which was great, on one hand, because nothing was known about it, so almost anything I did was new, but then I got to a point where I wanted to ask more probing, big-picture questions, and I couldn't do it on that species without tools that were not available [yet] or without doing decades' worth of background research." That's not what she wanted to spend decades doing, and so she began searching for a species whose social behavior, including its power dynamics, she could probe deeply into.

"Cichlids are so social," she continues, "and they are interesting from an evolutionary perspective, [with their] rapid adaptive changes, and I was really drawn to that model system."

Toward the end of her PhD program, Maruska was at a conference, presenting a poster, when Russ Fernald came over and started chatting with her. Fernald was legendary in the world of cichlid evolution and ecology. Maruska knew his work on neurobiology and hormones in the cichlid *Astatotilapia burtoni*, and he was already one of her targeted postdoc advisors. Soon they were exchanging emails, and not long after that, in 2007, Fernald offered her a postdoc position in his lab at Stanford University.

Maruska's work with Fernald was mostly on the behavior and neurobiology of reproduction in *A. burtoni*. High-ranking males dig pits in their territories to spawn, and when they are not chasing away other males, they are courting females. It is quite the sight. Males ramp up the intensity of their body colors and quiver and shake in front of females. As they do so, they also keep a lookout for subordinate males: "Even though subordinate males' testes are small, they are packed with sperm," Maruska notes. "They can still spawn with females. . . . If they get the chance, they can interject [themselves] between a spawning pair and they can try to fertilize some eggs." Assuming he doesn't have to chase off some such usurper, the dominant male produces pulsed courtship sounds and shakes his tail, turning to the pit and attempting to entice the female to follow.

Astatotilapia burtoni are mouthbrooders. If a pair begin to spawn, the female deposits her eggs in the male's pit. Then she picks up the eggs in her mouth, and while the male continues to display in front of her, he ejects sperm inside her mouth to fertilize the eggs. After it's all done, the female leaves with a mouth full of fertilized eggs, which she broods for two weeks before releasing her now free-swimming fry.

As wondrous as reproduction is to watch in this species, one can't study it without thinking about power. "The males are so territorial and their ability to control that territory dictates how successful they are reproductively," Maruska says. "If they don't have a territory that they are actively defending, they have very little chance of reproducing. The territory is it. And they need aggression to [keep it]." That got her thinking about power and, in particular, changes in power. A chance to delve into this topic opened when she got her own populations of A. burtoni up and running at Louisiana State University, where she had accepted an assistant professor position.

Maruska's lab houses more than a thousand A. burtoni. It's full of racks of shelves holding 30- to 50-gallon aquaria, each of which contains twenty fish, a mixture of males and females. Each tank is provided with two or three small flowerpots, and dominant males build their territories around those pots. Like virtually all laboratory populations of A. burtoni, Maruska's fish are descended from fish Fernald brought back from Lake Tanganyika in the 1970s. Maruska herself has never been to the lake: "Russ used to tell us that the region [where] these fish are commonly found has always been an area of political unrest," she says. "For [a] long time it wasn't safe to go there." But she's itching to go someday, when the time is right.

The more Maruska thought and read about power in cichlids, the more she realized that "nobody wants to go down in a hierarchy, [but] we know less about going down [in] a hierarchy than going up." She decided she needed to know more about going up and down the power ladder, both in terms of what happened behaviorally and what was driving the process from the hormonal, neurobiological, and genetic side.

Dominant, territorial A. burtoni males look and act very different from subordinate males. These powerful males have bright yellow and/or blue colors (subordinate males are drabber), a dark bar that runs through their eyes, a black spot on their gills, and a red patch just behind the gills. Subordinates lack all these characteristics. Dominant males spend their time defending their territories from other males and courting females, while subordinate males spend most of their time trying to avoid, or fleeing from, dominants. But, as Maruska and her colleagues know, when there is a power shake-up, and individuals ascend or descend in the hierarchy, all of that changes.[7]

To look at what happens when a fish loses its dominant status, Maruska and her team began by collecting males from the large communal tanks, taking only males that had a territory that they had aggressively defended for three consecutive days. Each of these males was then placed in a smaller experimental tank with two females and was given time to establish a new territory around a pot in that tank. Next, an opaque partition was lifted. Behind that partition was a very large male (10–20% bigger than the territory holder), a smaller male (5–10% smaller than the territory holder), or no fish at all. As Maruska and her colleagues had planned, the territory holders always lost control of their territory (and their dominant status) to a larger opponent, but always succeeded in defending their territory and rank against a smaller opponent.

Social descent was swift when territory holders were paired against larger intruders. Within thirty minutes, they had lost their eye bars and had muted their body colors. The number of aggressive behaviors they displayed plummeted equally quickly. Cortisol levels, a measure of stress, were more than double what they were in fish paired with small intruders. Perhaps most remarkably, Maruska's team found rapid changes in the expression of what are known as immediate early genes (IEGs): genes known to play a role in the nervous system's response to stimuli. Levels of expression of two of these genes, *cfos* and *egr-1*, were much higher in the brains of males after they had descended from power than in those of males that retained their territories. Exactly what those higher levels of *cfos* and *egr-1* expression translate into is a complex question that Maruska and others are investigating. What is clear is that losing status, territory, and, hence, power leads to massive behavioral, hormonal, neuronal, and gene-expression changes.[8]

Maruska and her colleagues have also mapped out changes in fish going up the power ladder. To do so, they began by placing a male in a communal tank with a larger, dominant individual. Then that male was placed in a smaller tank that contained a larger, territorial male and three or four females. Once Maruska had confirmed that the subject male had held subordinate status for two days, she or one of her team snuck into the lab at night, in the pitch-dark, wearing night-vision goggles, and removed the dominant male from the tank. Within minutes after the lights went on the next morning, the subject males were treating the territory as their own, and they were courting females. They also displayed more vibrant colors, turned on their eye bars, increased the size of their testes, ramped up their ability to produce sperm, and within thirty minutes had increased circulating levels of testosterone as well as various pituitary gland hormones. Maruska and her team hope one day to be able to strap on a "little recording transmitter that will do neural recordings as fish are swimming around." Should they succeed, it would be yet another potential tool available to use in the never-ending quest to better understand power in nonhuman societies.[9]

In the meantime, work on everything from the costs and benefits of power to assessment strategies used in the struggle for power; from winner, loser, bystander, and audience effects to the role of coalitions in acquiring and maintaining power; from interventions and groups challenging other groups for the reins of power to the rise and fall of the powerful continues full force in animals living everywhere on the planet.

AFTERWORD

It's an exciting time to be studying power in nonhuman societies. All around the globe, world-class scientists continue to peel back the layers and help us better understand the roots of power. Still, there is work to be done before we have, as we someday surely will, a truly integrative, conceptually based explanation for the dynamics of power in nonhumans.

Prognosticating what that explanation will look like is perilous, but it is fairly safe to say it will have certain general characteristics:

- It is bound to be complex because power is complex and multidimensional.
- It will, at its heart, be an evolutionary explanation, which is to say that it will involve an even deeper understanding, both at the empirical and the theoretical level, of the *costs and benefits of power*, as they play out in a *particular ecological context*, over *long periods of time*.
- It will be informed by new theory. My sense here is that this will involve a greater reliance on social network theory to better understand how the effects of power *move* through groups, and between groups, sometimes in a wavelike fashion.
- It will inform not just our understanding of power per se, but nonhuman behavior in general, for all power dynamics are embedded within the social milieu that makes up animal day-to-day life: feed-

ing, mating, raising young, habitat selection, protection, and so much more.

- It will be informed by new technological breakthroughs in data collection, some of which have only recently been adopted, and some of which are at the blueprint stage at most. There will never be a replacement for observing animals in the field (or lab) to test hypotheses about power. But how that is done is changing at a breakneck pace. We've seen some examples throughout the book: GPS tracking allows researchers not just to track capuchins and hyenas, but to make inferences about power dynamics; drones allow us to study how power unfolds in dolphins and will very soon in banded mongooses; underwater robots "see" in red light and inform us about the role of color change in the power struggles of cuttlefish.

  Much more is surely on the horizon. Animal behaviorists have already begun placing large-scale grids of video cameras, triggered by motion, in forests and elsewhere. To my knowledge, none of these grids have been put in place to study power per se, but the data that come from them will surely teach us things about power we have not even dreamt of. On an even grander scale, the Max Planck Institute of Animal Behavior (together with the Russian space agency and the German Aerospace Center) recently launched the Icarus satellite, dedicated to gathering data on large-scale animal migration patterns — data that are so fine-tuned that we can track particular individuals over vast spaces. Icarus transmitters, which weigh only about 5 grams (and soon will be reduced to 1 gram) are rolling off the assembly lines. In a pilot project, researchers have recently tagged five thousand blackbirds and thrushes in Eurasia, Russia, and the Americas to track their movements. I have no doubt but that clever ethologists will discover a way to use Icarus, or another satellite, to study power in non-human societies.

- It will be informed by breakthroughs in observing not only the locations of animals in space and how they interact with one other, but also what is going on inside animals as power dynamics play out. New advances in endocrinology and neurobiology will enable us to better grasp not only how behaviors associated with power affect hormone levels and neurobiological activity, but how hormone levels and neurobiological activity affect the dynamics of power, in part by tracking these effects in real time. In a similar vein, advances in the area of gene expression will help us understand how genes "turning off" and "turning on" impact the quest for power.

How all of this unfolds over the coming years will be fascinating to watch.

A closing thought: As I wrote this book, COVID-19 turned the world upside down. Social isolation became an all too necessary norm. Fortunately, all the scientists I write about throughout these pages took time away from their hectic schedules in the midst of a global pandemic to provide me with so much information, and so many wondrous tales, that I experienced a special comfort in vicariously living through their journeys as they studied power in nonhumans.

It is my sincere hope that you shared that same experience, that you could feel yourself looking over the stern of Richard Connor's boat watching dolphins sorting out their power structure in Shark Bay, Australia; sitting with Kate Holekamp over a campfire at the Fisi basecamp in Kenya sharing tales of hyenas and power; spending time in the idyllic Austrian Alps with Thomas Bugnyar watching raven power politics; and even lying there in the cave in New Zealand with Joe Waas as he pieced together little blue penguin power plays. Perhaps on an especially difficult day, it helped to imagine yourself studying power in street dogs in Kolkata, fairy wrens in Melbourne, elephant seals at Año Nuevo State Park in California, or white-fronted bee-eaters in Kenya.

Science and tales of adventure and wonder are a potent cocktail.

ACKNOWLEDGMENTS

I've long thought that my fellow animal behaviorists are, on average, the nicest lot of humans on the planet. Working on this book has only reinforced that belief. I'm indebted to all of the following researchers for allowing me to interview them *at length* (via Zoom, Skype, phone, and email) about their work on power in animals: Michael Alfieri, Tzo Zen Ang, Steven Austad, Cyrille Barrette, Matthew Bell, Anindita Bhadra, Roberto Bonanni, Elodie Briefer, Michelle Brown, Thomas Bugnyar, Michael Cant, Caroline Casey, Richard Connor, Meg Crofoot, Frans de Waal, Pedro Dias, Ryan Earley, Perri Eason, Robert Elwood, Steve Emlen, Magnus Enquist, Claudia Feh, John Fitzpatrick, Jessica Flack, Roger Hanlon, Kay Holekamp, David Holway, Domhnall Jennings, Konstanze Kruger, Burney Le Boeuf, Dario Maestripieri, Cathy Marler, Karen Maruska, John Mitani, Mark Moffett, Raoul Mulder, Craig Packer, Elisabetta Palagi, Walter Piper, Gordon Schuett, Michael Taborsky, Nahoko Tokuyama, Joseph Waas, and Klaus Zuberbühler.

Joe Calamia, my editor at the University of Chicago Press, provided thoughtful and insightful comments and suggestions, and his trust in me as a writer is sincerely appreciated.

I am fortunate to have family members who don't mind my barraging them with requests to read draft book chapters, and I am grateful to Dana and Aaron Dugatkin for providing insightful comments as this book gelled. Finally, I also thank my friend 2R, who knows a thing or two about power.

NOTES

Chapter One

1. E. Hemingway, *The Green Hills of Africa* (New York: Scribner, 1935). Sy Montgomery led me to this quote in her delightful book about Kay Holekamp, *The Hyena Scientist* (Boston: HMH Books for Young Readers, 2018).

2. K. E. Holekamp, B. Dantzer, G. Stricker, K. C. S. Yoshida, and S. Benson-Amram, Brains, brawn and sociality: A hyaena's tale, *Animal Behaviour* 103 (2015): 237–248; S. A. Wahaj, K. R. Guse, and K. E. Holekamp, Reconciliation in the spotted hyena (*Crocuta crocuta*), *Ethology* 107 (2001): 1057–1074; K. E. Holekamp and S. Benson-Amram, The evolution of intelligence in mammalian carnivores, *Interface Focus* 7 (2017), DOI 10.1098/rsfs.2016.0108; K. E. Holekamp, S. T. Sakai, and B. Lundrigan, Social intelligence in the spotted hyena (*Crocuta crocuta*), *Philosophical Transactions of the Royal Society of London* 362 (2007): 523–538.

3. H. Kruuk, *The Spotted Hyena: A Study of Predation and Social Behavior* (Chicago: University of Chicago Press, 1972).

4. H. E. Watts, J. B. Tanner, B. L. Lundrigan, and K. E. Holekamp, Post-weaning maternal effects and the evolution of female dominance in the spotted hyena, *Proceedings of the Royal Society of London* 276 (2009): 2291–2298.

5. K. E. Holekamp, L. Smale, and M. Szykman, Rank and reproduction in the female spotted hyaena, *Journal of Reproduction and Fertility* 108 (1996): 229–237; K. E. Holekamp and L. Smale, Dominance acquisition during mammalian social development: The inheritance of maternal rank, *American Zoologist* 31 (1991): 306–317; H. E. Watts, J. B. Tanner, B. L. Lundrigan, and K. E. Holekamp, Post-weaning maternal effects and the evolution of female dominance in the spotted hyena, *Proceedings of the Royal Society* 276 (2009): 2291–2298; Z. M. Laubach, C. D. Faulk, D. C. Dolinoy, L. Montrose, T. R. Jones, D. Ray, M. O. Pioon, and K. E. Holekamp, Early life social and ecological determinants of global DNA methylation in wild spotted hyenas, *Molecular Ecology* 28 (2019): 3799–3812, https://doi.org/10.1111/mec.15174.

6. B. J. Le Boeuf, P. Morris, and J. Reiter, Juvenile survivorship of north-ern elephant seals, in *Elephant Seals: Population Ecology, Behavior and Physiology*, B. J. Le Boeuf and R. Laws, eds. (Berkeley: University of California Press, 1994), 121–136.

7. For more on what Año Nuevo looks like, see www.parks.ca.gov/?page_id =523 and www.parks.ca.gov/?page_id=27613.

8. N. Mathevon, C. Casey, C. Reichmuth, and I. Charrier, Northern elephant seals memorize the rhythm and timbre of their rivals' voices, *Current Biology* 2 (2017): 2352–2356; C. Casey, I. Charrier, N. Mathevon, and C. Reichmuth, Rival assessment among northern elephant seals: Evidence of associative learn-ing during male-male contests, *Royal Society Open Science* 2 (2015), https://doi .org/10.1098/rsos.150228. For earlier work on vocalizations, see G. A. Bartho-lomew, The role of vocalization in the social behaviour of the northern elephant seal, *Animal Behaviour* 10 (1962): 7–14.

9. B. J. Le Boeuf and R. S. Peterson, Social status and mating activity in ele-phant seals, *Science* 163 (1969): 91–93; M. P. Haley, C. J. Deutsch, and B. J. Le Boeuf, Size, dominance and copulatory success in male northern elephant seals, *Mirounga angustirostris, Animal Behaviour* 48 (1994): 1249–1260.

10. B. Stewart, P. Yochem, H. Huber, R. DeLong, R. Jameson, W. Sydeman, S. Allen, and B. J. Le Boeuf, History and present status of the northern elephant seal population, in Le Boeuf and Laws, *Elephant Seals*, 29–48.

11. C. Cox and B. J. Le Boeuf, Female incitation of male competition: A mechanism in sexual selection, *American Naturalist* 111 (1977): 317–335.

12. S. l. Mesnick and B. J. Le Boeuf, Sexual-behavior of male northern ele-phant seals: 2, Female response to potentially injurious encounters, *Behaviour* 117 (1991): 262–280.

13. R. T. Hanlon and J. B. Messenger, Adaptive coloration in young cuttlefish (*Sepia officinalis*): The morphology and development of body patterns and their relation to behavior, *Philosophical Transactions of the Royal Society of London* 320 (1988): 437–487; R. T. Hanlon, M. J. Naud, J. W. Forsythe, K. Hall, A. C. Watson, and J. McKechnie, Adaptable night camouflage by cuttlefish, *American Natural-ist* 169 (2007): 543–551; J. J. Allen, L. M. Mathger, K. C. Buresch, T. Fetchko, M. Gardner, and R. T. Hanlon, Night vision by cuttlefish enables changeable cam-ouflage, *Journal of Experimental Biology* 213 (2010): 3953–3960; K. C. Buresch, K. M. Ulmer, D. Akkaynak, J. J. Allen, L. M. Mathger, M. Nakamura, and R. T. Hanlon, Cuttlefish adjust body pattern intensity with respect to substrate inten-sity to aid camouflage, but do not camouflage in extremely low light, *Journal of Experimental Marine Biology and Ecology* 462 (2015): 121–126; S. Zylinski, M. J. How, D. Osorio, R. T. Hanlon, and N. J. Marshall, To be seen or to hide: Visual characteristics of body patterns for camouflage and communication in the Aus-tralian giant cuttlefish *Sepia apama, American Naturalist* 177 (2011): 681–690; R. T. Hanlon, C. C. Chiao, L. M. Mathger, and N. J. Marshall, A fish-eye view of cuttlefish camouflage using in situ spectrometry, *Biological Journal of the Linnean Society* 109 (2013): 535–551. For more on swimming speed and respiration, see J. P. Aitken and R. K. O'Dor, Respirometry and swimming dynamics of the giant Australian cuttlefish, *Sepia apama* (Mollusca, Cephalopoda), *Marine and Fresh-water Behaviour and Physiology* 37 (2004): 217–234; N. L. Payne, B. M. Gillanders, R. S. Seymour, D. M. Webber, E. P. Snelling, and J. M. Semmens, Accelerometry

estimates field metabolic rate in giant Australian cuttlefish *Sepia apama* during breeding, *Journal of Animal Ecology* 80 (2011): 422–430.

14. J. J. Allen, G. R. R. Bell, A. M. Kuzirian, S. S. Velankar, and R. T. Hanlon, Comparative morphology of changeable skin papillae in octopus and cuttlefish, *Journal of Morphology* 275 (2014): 371–390; J. B. Messenger, Cephalopod chromatophores: Neurobiology and natural history, *Biological Reviews* 76 (2001): 473–528; C. C. Chiao, C. Chubb, and R. T. Hanlon, A review of visual perception mechanisms that regulate rapid adaptive camouflage in cuttlefish, *Journal of Comparative Physiology* A 201 (2015): 933–945.

15. K. C. Hall and R. T. Hanlon, Principal features of the mating system of a large spawning aggregation of the giant Australian cuttlefish *Sepia apama* (Mollusca: Cephalopoda), *Marine Biology* 140 (2002): 533–545.

16. For more on mating, see R. T. Hanlon, M. J. Naud, P. Shaw, and J. Havenhand, Transient sexual mimicry leads to fertilization, *Nature* 433 (2005): 212.

17. A. K. Schnell, C. L. Smith, R. T. Hanlon, and R. Harcourt, Giant Australian cuttlefish use mutual assessment to resolve male-male contests, *Animal Behaviour* 107 (2015): 31–40; A. K. Schnell, C. L. Smith, R. T. Hanlon, K. C. Hall, and R. Harcourt, Cuttlefish perform multiple agonistic displays to communicate a hierarchy of threats, *Behavioral Ecology and Sociobiology* 70 (2016): 1643–1655; A. K. Schnell, C. Jozet-Alves, K. C. Hall, L. Radday, and R. T. Hanlon, Fighting and mating success in giant Australian cuttlefish is influenced by behavioural lateralization, *Proceedings of the Royal Society of London* B 286 (2019), dx.doi .org/10.1098/rspb.2018.250.

18. R. P. Hannes and D. Franck, The effect of social isolation on androgen and corticosteroid levels in a cichlid fish (*Haplochromis burtoni*) and in swordtails (*Xiphophorus helleri*), *Hormones and Behavior* 17 (1983): 292–301; R. P. Hannes, D. Franck, and F. Liemann, Effects of rank-order fights on whole-body and blood concentration of androgen and corticosteroids in the male swordfish (*Xiphophorus helleri*), *Zeitschrift für Tierpsychologie* 65 (1984): 53–65; R. P. Hannes, Blood and whole-body androgen levels of male swordtails correlated with aggression measures in standard opponent test, *Aggressive Behavior* 12 (1986): 249–254; D. Franck and U. Wilhelmi, Changes of aggressive attack readiness of male swordfish, *Xiphophorus helleri*, after social isolation, *Experientia* 29 (1973): 896–897; D. Franck, The effect of social stimuli on steroid levels and attack readiness in male cichlids and swordtails, *Aggressive Behavior* 10 (1984): 154; A. Ribowski and D. Franck, Demonstration of strength and concealment of weakness in escalating fights of male swordtails (*Xiphophorus helleri*), *Ethology* 93 (1993): 265–274; A. Ribowski and D. Franck, Subordinate swordtail males escalate faster than dominants: A failure of the social conditioning principle, *Aggressive Behavior* 19 (1993): 223–229; D. Franck and A. Ribowski, Influences of prior agonistic experiences on aggression measures in the male swordtail (*Xiphophorus helleri*), *Behaviour* 103 (1987): 217–240.

19. D. Franck and A. Ribowski, Dominance hierarchies of male green swordtails in nature, *Journal of Fish Biology* 43 (1993): 497–499; D. Franck, B. Klamroth, A. Taebel-Hellwig, and M. Schartl, Home ranges and satellite tactics of male green swordtails (*Xiphophorus helleri*) in nature, *Behavioral Processes* 43 (1998): 115–123.

20. R. L. Earley, Aggression, Eavesdropping and Social Dynamics in Male

Green Swordtail Fish (*Xiphophorus helleri*) (PhD, University of Louisville, 2002); R. L. Earley and L. A. Dugatkin, Merging social hierarchies: Effects on dominance rank in male green swordtail fish (*Xiphophorus helleri*), *Behavioural Processes* 73 (2006): 290–298.

21. R. L. Earley and L. A. Dugatkin, Eavesdropping on visual cues in green swordtail (*Xiphophorus helleri*) fights: A case for networking, *Proceedings of the Royal Society of London* 269 (2002): 943–952; R. L. Earley, M. Druen, and L. A. Dugatkin, Watching fights does not alter a bystander's response towards naive conspecifics in male green swordtail fish, *Xiphophorus helleri*, *Animal Behaviour* 69 (2005): 1139–1145; R. L. Earley, M. Tinsley, and L. A. Dugatkin, To see or not to see: Does previewing a future opponent affect the contest behavior of green swordtail males (*Xiphophorus helleri*)? *Naturwissenschaften* 90 (2003): 226–230.

22. J. S. Lopes, R. Abril-de-Abreu, and R. F. Oliveira, Brain transcriptomic response to social eavesdropping in zebrafish (*Danio rerio*), *PLoS ONE* 10 (2015), https://doi.org/0.1371/journal.pone.0145801.

Chapter Two

1. T. Z. Ang and A. Manica, Aggression, segregation and stability in a dominance hierarchy, *Proceedings of the Royal Society of London* 277 (2010): 1337–1343.

2. T. Z. Ang and A. Manica, Benefits and costs of dominance in the angelfish *Centropyge bicolor*, *Ethology* 116 (2010): 855–865.

3. T. Z. Ang, Social Conflict Resolution in Groups of the Angelfish *Centropyge bicolor* (PhD, University of Cambridge, 2010), 87.

4. A. L. Martin and P. A. Moore, Field observations of agonism in the crayfish, *Orconectes rusticus*: Shelter use in a natural environment, *Ethology* 113 (2007): 1192–1201.

5. A. L. Martin and P. A. Moore, The influence of dominance on shelter preference and eviction rates in the crayfish, *Orconectes rusticus*, *Ethology* 114 (2008): 351–360.

6. P. M. Kappeler and A. Dill, The lemurs of Kirindy, *Natural History* 109 (2000): 58–65; D. Clough, M. Heistermann, and P. M. Kappeler, Host intrinsic determinants and potential consequences of parasite infection in free-ranging red-fronted lemurs (*Eulemur fulvus rufus*), *American Journal of Physical Anthropology* 142 (2010): 441–452; M. E. Pereira and P. M. Kappeler, Divergent systems of agonistic behaviour in lemurid primates, *Behaviour* 134 (1997): 225–274.

7. B. Habig and E. A. Archie, Social status, immune response and parasitism in males: A meta-analysis, *Philosophical Transactions of the Royal Society of London* 370 (2015), https://doi.org/10.1098/rstb.2014.0109; B. Habig, M. M. Doellman, K. Woods, J. Olansen, and A. Archie, Social status and parasitism in male and female vertebrates: A meta-analysis, *Scientific Reports* 8 (2018), https://doi.org /10.1038/s41598-018-21994-7.

8. M. S. Mooring, A. A. McKenzie, and B. L. Hart, Role of sex and breeding status in grooming and total tick load of impala, *Behavioral Ecology and Sociobiology* 39 (1996): 259–266; K. A. Lee, Linking immune defenses and life history at the levels of the individual and the species, *Integrative and Comparative Biology* 46 (2006): 1000–1015.

9. G. L. MacLean, The sociable weaver, part 2: Nest architecture and social organization, *Ostrich* 44 (1973): 191–218; E. C. Collias and N. E. Collias, Nest building and nesting-behavior of sociable weaver *Philetairus socius*, *Ibis* 120 (1978): 1–15; G. M. Leighton and L. Vander Meiden, Sociable weavers increase cooperative nest construction after suffering aggression, *PLoS ONE* 11 (2016), https://doi.org/10.1371/journal.pone.0150953; M. Paquet, C. Doutrelant, M. Loubon, F. Theron, M. Rat, and R. Covas, Communal roosting, thermoregulatory benefits and breeding group size predictability in cooperatively breeding sociable weavers, *Journal of Avian Biology* 47 (2016): 749–755; M. Rat, R. E. van Dijk, R. Covas, and C. Doutrelant, Dominance hierarchies and associated signalling in a cooperative passerine, *Behavioral Ecology and Sociobiology* 69 (2015): 437–448.

10. L. R. Silva, S. Lardy, A. C. Ferreira, B. Rey, C. Doutrelant, and R. Covas, Females pay the oxidative cost of dominance in a highly social bird, *Animal Behaviour* 144 (2018): 135–146; A. Cohen, K. Klasing, and R. Ricklefs, Measuring circulating antioxidants in wild birds, *Comparative Biochemistry and Physiology* B: *Biochemistry and Molecular Biology* 147 (2007): 110–121; T. Finkel and N. J. Holbrook, Oxidants, oxidative stress and the biology of ageing, *Nature* 408 (2000): 239–247.

11. M. B. V. Bell, M. A. Cant, C. Borgeaud, N. Thavarajah, J. Samson, and T. H. Clutton-Brock, Suppressing subordinate reproduction provides benefits to dominants in cooperative societies of meerkats, *Nature Communications* 5 (2014), https://doi.org/10.1038/ncomms5499.

12. A. J. Young, A. A. Carlson, S. L. Monfort, A. F. Russell, N. C. Bennett, and T. H. Clutton-Brock, Stress and the suppression of subordinate reproduction in cooperatively breeding meerkats, *Proceedings of the National Academy of Sciences* 103 (2006): 12005–12010; A. J. Young and T. H. Clutton-Brock, Infanticide by subordinates influences reproductive sharing in cooperatively breeding meerkats, *Biology Letters* 2 (2006): 385–387.

13. K. N. Smyth, N. M. Caruso, C. S. Davies, T. H. Clutton-Brock, and C. M. Drea, Social and endocrine correlates of immune function in meerkats: Implications for the immunocompetence handicap hypothesis, *Royal Society Open Science* 5 (2018), https://doi.org/10.1098/rsos.180435.

14. Cornell University has some excellent online videos on the behavior of common loons: https://www.cornell.edu/video/loon-territoriality.

15. Larger lakes sometimes house numerous pairs, each of which have cordoned off a home of their own. W. Piper, J. N. Mager, and C. Walcott, Marking loons, making progress: Striking discoveries about the social behavior and communication of common loons are revealed by a low-tech approach, *American Scientist* 99 (2011): 220–227; J. N. Mager, C. Walcott, and W. H. Piper, Male common loons signal greater aggressive motivation by lengthening territorial yodels, *Wilson Journal of Ornithology* 124 (2012): 73–80; J. N. Mager and C. Walcott, Dynamics of an aggressive vocalization in the common loon (*Gavia immer*): A review, *Waterbirds* 37 (2014): 37–46.

16. W. H. Piper, K. M. Brunk, G. L. Jukkala, E. A. Andrews, S. R. Yund, and N. G. Gould, Aging male loons make a terminal investment in territory defense, *Behavioral Ecology and Sociobiology* 72 (2018), https://doi.org/10.1007/s00265-018-2511-9.

17. W. H. Piper, C. Walcott, J. N. Mager, and F. J. Spilker, Fatal battles in common loons: A preliminary analysis, *Animal Behaviour* 75 (2008): 1109–1115; W. H. Piper, J. N. Mager, C. Walcott, L. Furey, N. Banfield, A. Reinke, F. J. Spilker, and J. A. Flory, Territory settlement in common loons: No footholds but age and assessment are important, *Animal Behaviour* 104 (2015): 155–163; W. H. Piper, K. M. Brunk, J. A. Flory, and M. W. Meyer, The long shadow of senescence: Age impacts survival and territory defense in loons, *Journal of Avian Biology* 48 (2017): 1062–1070; J. A. Spool, L. V. Riters, and W. H. Piper, Investment in territorial defence relates to recent reproductive success in common loons *Gavia immer*, *Journal of Avian Biology* 48 (2017): 1281–1286; J. N. Mager and C. Walcott, Dynamics of an aggressive vocalization in the common loon (*Gavia immer*): A review, *Waterbirds* 3 (2014): 37–46.

Chapter Three

1. T. H. Huxley (1888), The struggle for existence: A programme, *Nineteenth Century* 23, 161–180; T. H. Huxley, *Collected Essays*, vol. 1 (New York: Macmillan, 1893); Thomas Henry Huxley to Charles Darwin, November 23, 1859, Darwin Correspondence Project, https://www.darwinproject.ac.uk/letter/DCP-LETT -2544.xml.

2. Charles Darwin to T. H. Huxley, August 8, 1860, Darwin Correspondence Project, https://www.darwinproject.ac.uk/letter/DCP-LETT-2893.xml; T. Huxley, *Autobiography and Selected Essays* (Boston: Houghton-Mifflin, 1909), 13.

3. For more on the Grands-Jardins National Park, see D. Vandal, Le caribou des Grands-jardins: Légende et réalité, *Carnets de Zoologie* 43 (1983): 36–41, https://www.sepaq.com/pq/grj/index.dot.

4. D. I. Chapman, Antlers: Bones of contention, *Mammal Review* 5 (1975): 121–172; R. J. Goss, *Deer Antlers: Regeneration, Function and Evolution* (Cambridge, MA: Academic Press, 1983).

5. C. Barrette and D. Vandal, Sparring, relative antler size, and assessment in male caribou, *Behavioral Ecology and Sociobiology* 26 (1990): 383–387.

6. C. Barrette and D. Vandal, Sparring and access to food in female caribou in the winter, *Animal Behaviour* 40 (1990): 1183–1185; C. Barrette and D. Vandal, Social rank, dominance, antler size, and access to food in snow-bound wild woodland caribou, *Behaviour* 97 (1986): 118–146; C. Barrette and D. Vandal, Sparring, relative antler size, and assessment in male caribou, *Behavioral Ecology and Sociobiology* 26 (1990): 383–387.

7. Z. S. Lin, L. Chen, X. Q. Chen, Y. B. Zhong, Y. Yang, W. H. Xia, C. Liu et al., Biological adaptations in the Arctic cervid, the reindeer (*Rangifer tarandus*), *Science* 364 (2019), https://doi.org/10.1126/science.aav6312; Z. P. Li, Z. S. Lin, H. X. Ba, L. Chen, Y. Z. Yang, K. Wang, Q. Qiu, W. Wang, and G. Y. Li, Draft genome of the reindeer (*Rangifer tarandus*), *Gigascience* 6 (2017), https://doi.org/10.1093 /gigascience/gix102.

8. C. Barrette and D. Vandal, Sparring and access to food in female caribou in the winter, *Animal Behaviour* 40 (1990): 1183–1185. For more on antlers in females, see L. E. Loe, G. Pigeon, S. D. Albon, P. E. Giske, R. J. Irvine, E. Ropstad, A. Stien, V. Veiberg, and A. Mysterud, Antler growth as a cost of reproduc-

tion in female reindeer, *Oecologia* 189 (2019): 601–609; J. A. Schaefer and S. P. Mahoney, Antlers on female caribou: Biogeography of the bones of contention, *Ecology* 82 (2001): 3556–3560; L. Gagnon and C. Barrette, Antler casting and parturition in wild female caribou, *Journal of Mammalogy* 73 (1992): 440–442.

9. R. W. Elwood and S. Neil, *Assessments and Decisions: A Study of Information Gathering by Hermit Crabs* (London: Chapman and Hall, 1992); R. W. Elwood and M. Briffa, Information gathering and communication during agonistic encounters: A case study of hermit crabs, *Advances in the Study of Behavior* 30 (2001): 53–97; R. W. Elwood, N. Marks, and J. Dick, Consequences of shell species preferences for female reproductive success in the hermit crab *Pagurus bernhardus*, *Marine Biology* 123 (1995): 431–434.

10. R. W. Elwood, A. McClean, and L. Webb, Development of shell preferences by the hermit crab *Pagurus bernhardus*, *Animal Behaviour* 27 (1979): 940–946; R. W. Elwood, Motivational change during resource assessment by hermit crabs, *Journal of Experimental Marine Biology and Ecology* 193 (1995): 41–55.

11. B. A. Hazlett, Shell exchanges in hermit crabs: Aggression, negotiation, or both? *Animal Behaviour* 26 (1978): 1278–1279; R. W. Elwood and A. Stewart, The timing of decisions during shell investigation by the hermit crab, *Pagurus bernhardus*, *Animal Behaviour* 33 (1985): 620–627; R. W. Elwood, R. M. E. Pothanikat, and M. Briffa, Honest and dishonest displays, motivational state and subsequent decisions in hermit crab shell fights, *Animal Behaviour* 72 (2006): 853–859; M. Briffa and R. W. Elwood, The power of shell rapping influences rates of eviction in hermit crabs, *Behavioral Ecology* 11 (2000): 288–293; M. Briffa, R. W. Elwood, and M. Russ, Analysis of multiple aspects of a repeated signal: Power and rate of rapping during shell fights in hermit crabs, *Behavioral Ecology* 14 (2003): 74–79.

12. B. M. Dowds and R. W. Elwood, Shell wars: Assessment strategies and the timing of decisions in hermit crab shell fights, *Behaviour* 85 (1983): 1–24; B. M. Dowds and R. W. Elwood, Shell wars 2: The influence of relative size on decisions made during hermit crab shell fights, *Animal Behaviour* 33 (1985): 649–656.

13. M. Briffa and R. W. Elwood, Rapid change in energy status in fighting animals: Causes and effects of strategic decisions, *Animal Behaviour* 70 (2005): 119–124; S. M. Lane and M. Briffa, The role of spatial accuracy and precision in hermit crab contests, *Animal Behaviour* 167 (2020): 111–118.

14. M. Enquist and O. Leimar, Evolution of fighting behaviour: Decision rules and assessment of relative strength, *Journal of Theoretical Biology* 102 (1983): 387–410; O. Leimar and M. Enquist, Effects of asymmetries in owner intruder conflicts, *Journal of Theoretical Biology* 111 (1984): 475–491; M. Enquist and O. Leimar, Evolution of fighting behavior: The effect of variation in resource value, *Journal of Theoretical Biology* 127 (1987): 187–205.

15. M. Enquist, O. Leimar, T. Ljungberg, Y. Mallner, and N. Segardahl, A test of the sequential assessment game: Fighting in the cichlid fish, *Nannacara anomala*, *Animal Behaviour* 40 (1990): 1–15; M. Enquist, T. Ljungberg, and A. Zandor, Visual assessment of fighting ability in the cichlid fish *Nannacara anomala*, *Animal Behaviour* 35 (1987): 1262–1263.

16. S. N. Austad, A game theoretical interpretation of male combat in the

bowl and doily spider, *Animal Behaviour* 31 (1983): 59–73; O. Leimar, S. Austad, and M. Enquist, A test of the sequential assessment game: Fighting in the bowl and doily spider *Frontinella pyramitela*, *Evolution* 45 (1991): 862–874.

17. P. K. Eason, G. A. Cobbs, and K. G. Trinca, The use of landmarks to define territorial boundaries, *Animal Behaviour* 58 (1999): 85–91. For a model of landmarks and territoriality, see M. Mesterton-Gibbons and E. S. Adams, Landmarks in territory partitioning: A strategically stable convention? *American Naturalist* 161 (2003): 685–697.

18. J. R. LaManna and P. K. Eason, Effects of landmarks on territorial establishment, *Animal Behaviour* 65 (2003): 471–478; P. S. Suriyampola and P. K. Eason, A field study investigating effects of landmarks on territory size and shape, *Biology Letters* 10 (2014), https://doi.org/10.1098/rsbl.2014.0009; P. S. Suriyampola and P. K. Eason, The effects of landmarks on territorial behavior in a convict cichlid, *Amatitlania siquia*, *Ethology* 121 (2015): 785–792; E. S. Adams, Approaches to the study of territory size and shape, *Annual Review of Ecology and Systematics* 32 (2001): 277–303; M. Andersson, Optimal foraging area: Size and allocation of search effort, *Theoretical Population Biology* 13 (1978): 397–409.

19. Fisher, J., Evolution and bird sociality, in *Evolution as a Process*, J. Huxley, A. C. Hardy, and E. B. Ford, eds. (Sydney: Allen and Unwin, 1954), 73; R. Jaeger, Dear enemy recognition and the costs of aggression between salamanders, *American Naturalist* 117 (1981): 962–974; T. Getty, Dear enemies and the prisoners-dilemma: Why should territorial neighbors form defensive coalitions, *American Zoologist* 27 (1987): 327–336.

20. E. Briefer, T. Aubin, K. Lehongre, and F. Rybak, How to identify dear enemies: The group signature in the complex song of the skylark *Alauda arvensis*, *Journal of Experimental Biology* 211 (2008): 317–326; E. Briefer, F. Rybak, and T. Aubin, When to be a dear enemy: Flexible acoustic relationships of neighbouring skylarks, *Alauda arvensis*, *Animal Behaviour* 76 (2008): 1319–1325; E. Briefer, T. Aubin, and F. Rybak, Response to displaced neighbours in a territorial songbird with a large repertoire, *Naturwissenschaften* 96 (2009): 1067–1077; E. Briefer, F. Rybak, and T. Aubin, Are unfamiliar neighbours considered to be dear-enemies? *PLoS ONE* 5 (2010), https://doi.org/10.1371/journal.pone .0012428.

Chapter Four

1. G. Szipl, E. Ringler, M. Spreafico, and T. Bugnyar, Calls during agonistic interactions vary with arousal and raise audience attention in ravens, *Frontiers in Zoology* 14 (2017), https://doi.org/10.1186/s12983-017-0244-7; J. J. M. Massen, A. Pasukonis, J. Schmidt, and T. Bugnyar, Ravens notice dominance reversals among conspecifics within and outside their social group, *Nature Communications* 5 (2014), https://doi.org/10.1038/ncomms4679; B. Heinrich, *Ravens in Winter* (New York: Summit Books, 1989).

2. G. Szipl, E. Ringler, and T. Bugnyar, Attacked ravens flexibly adjust signalling behaviour according to audience composition, *Proceedings of the Royal Society of London* 285 (2018), https://doi.org/10.1098/rspb.2018.0375.

3. K. Zuberbühler, Referential labelling in Diana monkeys, *Animal Behaviour*

59 (2000): 917–927; K. Zuberbühler, Causal knowledge of predators' behaviour in wild Diana monkeys, *Animal Behaviour* 59 (2000): 209–220.

4. K. E. Slocombe and K. Zuberbühler, Chimpanzees modify recruitment screams as a function of audience composition, *Proceedings of the National Academy of Sciences* 104 (2007): 17228–17233.

5. L. C. dos Santos, F. A. D. Freire, and A. C. Luchiari, The effect of audience on intrasexual interaction in the male fiddler crab, *Uca maracoani, Journal of Ethology* 35 (2017): 93–100; S. K. Darden, M. K. May, N. K. Boyland, and T. Dabelsteen, Territorial defense in a network: Audiences only matter to male fiddler crabs primed for confrontation, *Behavioral Ecology* 30 (2019): 336–340; K. Hirschenhauser, M. Gahr, and W. Goymann, Winning and losing in public: Audiences direct future success in Japanese quail, *Hormones and Behavior* 63 (2013): 625–633; R. J. Matos, T. M. Peake, and P. K. McGregor, Timing of presentation of an audience: Aggressive priming and audience effects in male displays of Siamese fighting fish (*Betta splendens*), *Behavioural Processes* 63 (2003): 53–61; T. L. Dzieweczynski, R. L. Earley, T. M. Green, and W. J. Rowland, Audience effect is context dependent in Siamese fighting fish, *Betta splendens, Behavioral Ecology* 16 (2005): 1025–1030. T. L. Dzieweczynski, A. C. Eklund, and W. J. Rowland, Male 11-ketotestosterone levels change as a result of being watched in Siamese fighting fish, *Betta splendens, General and Comparative Endocrinology* 147 (2006): 184–189; T. L. Dzieweczynski and C. E. Perazio, I know you: Familiarity with an audience influences male-male interactions in Siamese fighting fish, *Betta splendens, Behavioral Ecology and Sociobiology* 66 (2012): 1277–1284; T. L. Dzieweczynski, C. E. Gill, and C. E. Perazio, Opponent familiarity influences the audience effect in male-male interactions in Siamese fighting fish, *Animal Behaviour* 83 (2012): 1219–1224; F. Bertucci, R. J. Matos, and T. Dabelsteen, Knowing your audience affects male-male interactions in Siamese fighting fish (*Betta splendens*), *Animal Cognition* 17 (2014): 229–236.

6. For a review of the theoretical work on winner and loser effects, see W. Lindquist and I. D. Chase, Data-based analysis of winner-loser models of hierarchy formation in animals, *Bulletin of Mathematical Biology* 71 (2009): 556–584; and M. Mesterton-Gibbons, Y. Dai, and M. Goubault, Modeling the evolution of winner and loser effects: A survey and prospectus, *Mathematical Biosciences* 274 (2016): 33–44.

7. G. W. Schuett and J. C. Gillingham, Male-male agonistic behavior of the copperhead, *Agkistrodon contortrix, Amphibia-Reptilia* 10 (1989): 243–266.

8. G. W. Schuett, Body size and agonistic experience affect dominance and mating success in male copperheads, *Animal Behaviour* 54 (1997): 213–224; N. Angier, Pit viper's life: Bizarre, gallant and venomous, *New York Times*, October 15, 1991.

9. G. W, Schuett, H. J. Harlow, J. D. Rose, J., E. A. Van Kirk, and W. J. Murdoch, Levels of plasma corticosterone and testosterone in male copperheads (*Agkistrodon contortrix*) following staged fights, *Hormones and Behavior* 30 (1996): 60–68; G. W. Schuett and M. S. Grober, Post-fight levels of plasma lactate and corticosterone in male copperheads, *Agkistrodon contortrix* (Serpentes, Viperidae): Differences between winners and losers, *Physiology and Behavior* 71 (2000): 335–341. Exactly how all this works remains unclear, but Schuett and

his team hypothesize that high levels of plasma corticosterone shunt lactate, which converts food to energy, from muscles (where it would be especially useful for aggressive interactions) to blood.

10. M. Mesterton-Gibbons, On the evolution of pure winner and loser effects: A game-theoretic model, *Bulletin of Mathematical Biology* 61 (1999): 1151–1186; M. Mesterton-Gibbons, Y. Dai, and M. Goubault, Modeling the evolution of winner and loser effects: A survey and prospectus, *Mathematical Biosciences* 274 (2016): 33–44.

11. J. K. Bester-Meredith, L. J. Young, and C. A. Marler, Species differences in paternal behavior and aggression in *Peromyscus* and their associations with vasopressin immunoreactivity and receptors, *Hormones and Behavior* 36 (1999): 25–38; J. I. Terranova, C. F. Ferris, and H. E. Albers, Sex differences in the regulation of offensive aggression and dominance by Arginine-Vasopressin, *Frontiers in Endocrinology* 8 (2017), https://doi.org/10.3389/fendo.2017.0030. Prairie voles (*Microtus ochrogaster*) and meadow voles (*Microtus pennsylvanicus*) are another set of closely related species that differ in social structure: Z. R. Donaldson and L. J. Young, Oxytocin, vasopressin, and the neurogenetics of sociality, *Science* 322 (2008): 900–904; M. M. Lim, Z. X. Wang, D. E. Olazabal, X. H. Ren, E. F. Terwilliger, and L. J. Young, Enhanced partner preference in a promiscuous species by manipulating the expression of a single gene, *Nature* 429 (2004): 754–757; L. J. Young and Z. X. Wang, The neurobiology of pair bonding, *Nature Neuroscience* 7 (2004): 1048–1054; T. R. Insel and L. J. Young, The neurobiology of attachment, *Nature Reviews Neuroscience* 2 (2001): 129–136.

12. J. K. Bester-Meredith and C. A. Marler, Vasopressin and aggression in cross-fostered California mice (*Peromyscus californicus*) and white-footed mice (*Peromyscus leucopus*), *Hormones and Behavior* 40 (2001): 51–64; For more on cross-fostering, see L. A. Dugatkin, *Principles of Animal Behavior*, 4th ed. (Chicago: University of Chicago Press, 2020).

13. M. J. Fuxjager, G. Mast, E. A. Becker, and C. A. Marler, The "home advantage" is necessary for a full winner effect and changes in post-encounter testosterone, *Hormones and Behavior* 56 (2009): 214–219; M. J. Fuxjager, J. L. Montgomery, E. A. Becker, and C. A. Marler, Deciding to win: Interactive effects of residency, resources and "boldness" on contest outcome in white-footed mice, *Animal Behaviour* 80 (2010): 921–927; M. J. Fuxjager and C. A. Marler, How and why the winner effect forms: Influences of contest environment and species differences, *Behavioral Ecology* 21 (2010): 37–45; M. J. Fuxjager, T. O. Oyegbile, and C. A. Marler, Independent and additive contributions of post-victory testosterone and social experience to the development of the winner effect, *Endocrinology* 152 (2011): 3422–3429, https://doi.org/10.1210/en.2011-1099; M. J. Fuxjager, R. M. Forbes-Lorman, D. J. Coss, C. J. Auger, A. P. Auger, and C. A. Marler, Winning territorial disputes selectively enhances androgen sensitivity in neural pathways related to motivation and social aggression, *Proceedings of the National Academy of Sciences* 107 (2010): 12393–12398; E. A. Becker and C. A. Marler, Postcontest blockade of dopamine receptors inhibits development of the winner effect in the California mouse (*Peromyscus californicus*), *Behavioral Neuroscience* 129 (2015): 205–213. It's unclear yet why males with two prior wins had the same testosterone surge but were no more likely to win.

14. M. J. Fuxjager, J. L. Montgomery, and C. A. Marler, Species differences in

the winner effect disappear in response to post-victory testosterone manipulations, *Proceedings of the Royal Society of London* 278 (2011): 3497–3503.

15. K. Lorenz, The triumph ceremony of the greylag goose, *Philosophical Transactions of the Royal Society of London* 251 (1965): 477–481; J. R. Waas, Acoustic displays facilitate courtship in little blue penguins, *Eudyptula minor*, *Animal Behaviour* 36 (1988): 366–371; M. Miyazaki and J. R. Waas, Acoustic properties of male advertisement and their impact on female responsiveness in little penguins *Eudyptula minor*, *Journal of Avian Biology* 34 (2003): 229–232.

16. S. C. Mouterde, D. M. Duganzich, L. E. Molles, S. Helps, F. Helps, and J. R. Waas, Triumph displays inform eavesdropping little blue penguins of new dominance asymmetries, *Animal Behaviour* 83 (2012): 605–611.

Chapter Five

1. N. Humphrey, The social function of intellect, in *Growing Points in Ethology*, P. Bateson and R. Hinde, eds. (Cambridge: Cambridge University Press, 1976), 303–317. For more on primate social intelligence, see S. M. Reader, Y. Hager, and K. N. Laland, The evolution of primate general and cultural intelligence, *Philosophical Transactions of the Royal Society of London* 366 (2011): 1017–1027; R. W. Byrne and A. Whiten (eds.), *Machiavellian Intelligence: Social Expertise and the Evolution of Intellect in Monkeys, Apes and Humans* (Oxford: Oxford University Press, 1988); R. I. Dunbar, The social brain: Mind, language and society in evolutionary perspective, *Annual Review of Anthropology* 325 (2003): 163–181; D. L. Cheney and R. M. Seyfarth, *Baboon Metaphysics: The Evolution of a Social Mind* (Chicago: University of Chicago Press, 2007); F. B. M. de Waal and P. Tyack (eds.), *Animal Social Complexity* (Chicago: University of Chicago Press, 2003); M. Tomasello and J. Call, *Primate Cognition* (Oxford: Oxford University Press, 1997).

2. K. E. Holekamp, B. Dantzer, G. Stricker, K. C. S. Yoshida, and S. Benson-Amram, Brains, brawn and sociality: A hyaena's tale, *Animal Behaviour* 103 (2015): 237–248; K. E. Holekamp and S. Benson-Amram, The evolution of intelligence in mammalian carnivores, *Interface Focus* 7 (2017), https://doi .org/10.1098/rsfs.2016.0108; S. Benson-Amram, B. Dantzer, G. Strickere, E. Swanson, and K. Holekamp, Brain size predicts problem-solving ability in mammalian carnivores, *Proceedings of the National Academy of Sciences* 113 (2016): 2532–2537; J. E. Smith, R. C. Van Horn, K. S. Powning, A. R. Cole, K. E. Graham, S. K. Memenis, and K. E. Holekamp, Evolutionary forces favoring intragroup coalitions among spotted hyenas and other animals, *Behavioral Ecology* 21 (2010): 284–303; E. D. Strauss and K. E. Holekamp, Social alliances improve rank and fitness in convention-based societies, *Proceedings of the National Academy of Sciences* 116 (2019): 8919–8924.

3. Coalition size was usually two, but, on occasion, could be larger. For more on inclusive fitness theory, see W. D. Hamilton, The evolution of altruistic behavior, *American Naturalist* 97 (1963): 354–356; W. D. Hamilton, The genetical evolution of social behaviour, I and II, *Journal of Theoretical Biology* 7 (1964): 1–52; and L. A. Dugatkin, *The Altruism Equation: Seven Scientists Search for the Origins of Goodness* (Princeton: Princeton University Press, 2006).

4. J. J. M. Massen, A. Pasukonis, J. Schmidt, and T. Bugnyar, Ravens notice

dominance reversals among conspecifics within and outside their social group, *Nature Communications* 5 (2014), https://doi.org/10.1038/ncomms4679; M. Boeckle and T. Bugnyar, Long-term memory for affiliates in ravens, *Current Biology* 22 (2012): 801–806; J. J. M. Massen, G. Szipl, M. Spreafico, and T. Bugnyar, Ravens intervene in others' bonding attempts, *Current Biology* 24 (2014): 2733–2736.

5. For reviews of Connor's work on power and coalitions in bottlenose dolphins, see R. C. Connor, *Dolphin Politics in Shark Bay* (New Bedford, MA: The Dolphin Alliance Project, 2018); and R. C. Connor and M. Krützen, Male dolphin alliances in Shark Bay: Changing perspectives in a 30-year study, *Animal Behaviour* 103 (2015): 223–235.

6. R. C. Connor, R. A. Smolker, and A. F. Richards, Two levels of alliance formation among male bottleneck dolphins, *Proceedings of the National Academy of Sciences* 89 (1992): 987–990; M. M. Wallen, E. M. Patterson, E. Krzyszczyk, and J. Mann, The ecological costs to females in a system with allied sexual coercion, *Animal Behaviour* 115 (2016): 227–236; R. C. Connor, Complex alliance relationships in bottlenose dolphins and a consideration of selective environments for extreme brain size evolution in mammals, *Philosophical Transactions of the Royal Society of London* 362 (2007): 587–602; M. Krützen, W. B. Sherwin, R. C. Connor, L. M. Barré, T. Van de Casteele, J. Mann, and R. Brooks, Contrasting relatedness patterns in bottlenose dolphins (*Tursiops* sp.) with different alliance strategies, *Proceedings of the Royal Society of London* 270 (2003): 497–502; R. C. Connor, R. Smolker, and L. Bejder, Synchrony, social behaviour and alliance affiliation in Indian Ocean bottlenose dolphins, *Tursiops aduncus*, *Animal Behaviour* 72 (2006): 1371–1378; C. H. Frere, M. Krützen, J. Mann, R. C. Connor, L. Bejder, and W. B. Sherwin, Social and genetic interactions drive fitness variation in a free-living dolphin population, *Proceedings of the National Academy of Sciences* 107 (2010): 19949–19954; L. M. Moller, L. B. Beheregaray, R. G. Harcourt, and M. Krützen, Alliance membership and kinship in wild male bottlenose dolphins (*Tursiops aduncus*) of southeastern Australia, *Proceedings of the Royal Society of London* 268 (2001): 1941–1947.

7. C. Feh, Alliances and reproductive success in Camargue stallions, *Animal Behaviour* 57 (1999): 705–713; C. Feh, Long term paternity data in relation to different aspects of rank for Camargue stallions, *Equus caballus*, *Animal Behaviour* 40 (1990): 995–996.

8. P. Ehrlich, *The Population Bomb* (New York: Ballantine Books, 1968).

9. R. L. Trivers, The evolution of reciprocal altruism, *Quarterly Review of Biology* 46 (1971): 189–226; C. Packer, Reciprocal altruism in *Papio anubis*, *Nature* 265 (1977): 441–443.

10. M. Mesterton-Gibbons and T. Sherratt, Coalition formation: A game-theoretical analysis, *Behavioral Ecology* 18 (2007): 277–286; M. Mesterton-Gibbons, S. Gavrilets, J. Gravner, and E. Akcay, Models of coalition or alliance formation, *Journal of Theoretical Biology* 274 (2011): 187–204. For more on theory and coalitions, see H. Whitehead and R. Connor, Alliances I. How large should alliances be? *Animal Behaviour* 69 (2005): 117–126; A. Bissonnette, S. Perry, L. Barrett, J. C. Mitani, M. Flinn, S. Gavrilets, and F. B. M. de Waal, Coalitions in theory and reality: A review of pertinent variables and processes, *Behav-*

iour 152 (2015): 1–56; R. A. Johnstone and L. A. Dugatkin, Coalition formation in animals and the nature of winner and loser effects, *Proceedings of the Royal Society of London* 267 (2000): 17–21; and M. Broom, A. Koenig, and C. Borries, Variation in dominance hierarchies among group-living animals: Modeling stability and the likelihood of coalitions, *Behavioral Ecology* 20 (2009): 844–855.

11. F. B. M. de Waal, *Chimpanzee Politics* (Baltimore: Johns Hopkins University Press, 1982).

12. F. B. M. de Waal, Sex differences in the formation of coalitions among chimpanzees, *Ethology and Sociobiology* 5 (1984): 239–255; F. B. M. de Waal, The integration of dominance and social bonding in primates, *Quarterly Review of Biology* 61 (1986): 459–479; F. B. M. de Waal, Coalitions as part of reciprocal relations in the Arnhem chimpanzee colony, in *Coalitions and Alliances in Humans and Other Animals*, A. Harcourt and F. B. M. de Waal, eds. (Oxford: Oxford University Press, 1992), 233–257.

13. N. Tokuyama and T. Furuichi, Do friends help each other? Patterns of female coalition formation in wild bonobos at Wamba, *Animal Behaviour* 119 (2016): 27–35; N. Angier, In the bonobo world, female camaraderie prevails, *New York Times*, September 10, 2016.

Chapter Six

1. S. T. Emlen and P. H. Wrege, Parent-offspring conflict and the recruitment of helpers among bee-eaters, *Nature* 356 (1992): 331–333.

2. T. H. Clutton-Brock, F. E. Guinness, and S. D. Albon, *Red Deer: The Behavior and Ecology of Two Sexes* (Chicago: University of Chicago Press, 1982); D. J. Jennings, M. P. Gammell, C. M. Carlin, and T. J. Hayden, Effect of body weight, antler length, resource value and experience on fight duration and intensity in fallow deer, *Animal Behaviour* 68 (2004): 213–221; D. J. Jennings, M. P. Gammell, C. M. Carlin, and T. J. Hayden, Win, lose or draw: A comparison of fight structure based on fight conclusion in the fallow deer, *Behaviour* 142 (2005): 423–439; D. J. Jennings, M. P. Gammell, C. M. Carlin, and T. J. Hayden, Is the parallel walk between competing male fallow deer, *Dama dama*, a lateral display of individual quality? *Animal Behaviour* 65 (2003): 1005–1012; D. J. Jennings, R. W. Elwood, T. J. Carlin, T. J. Hayden, and M. P. Gammell, Vocal rate as an assessment process during fallow deer contests, *Behavioural Processes* 91 (2012): 152–158; F. Alvarez, Risks of fighting in relation to age and territory holding in fallow deer, *Canadian Journal of Zoology* 71 (1993): 376–383; D. J. Jennings, R. J. Boys, and M. P. Gammell, Weapon damage is associated with contest dynamics but not mating success in fallow deer (*Dama dama*), *Biology Letters* 13 (2017), https://doi.org/10.1098/rsbl.2017.0565.

3. D. J. Jennings, R. J. Boys, and M. P. Gammell, Suffering third-party intervention during fighting is associated with reduced mating success in the fallow deer, *Animal Behaviour* 139 (2018): 1–8; D. J. Jennings, C. M. Carlin, and M. P. Gammell, A winner effect supports third-party intervention behaviour during fallow deer, *Dama dama*, fights, *Animal Behaviour* 77 (2009): 343–348; D. J. Jennings, C. M. Carlin, T. J. Hayden, and M. P. Gammell, Third-party intervention behaviour during fallow deer fights: The role of dominance, age, fighting and

body size, *Animal Behaviour* 81 (2011): 1217–1222; D. J. Jennings, R. J. Boys, and M. P. Gammell, Investigating variation in third-party intervention behavior during a fallow deer (*Dama dama*) rut, *Behavioral Ecology* 28 (2017): 288–293; L. A. Dugatkin, Breaking up fights between others: A model of intervention behaviour, *Proceedings of the Royal Society of London* 265 (1998): 443–437.

4. I. S. Bernstein, A field study of the pigtail monkey (*Macaca nemestrina*), *Primates* 8 (1967): 217–238; T. Oi, Patterns of dominance and affiliation in wild pig-tailed macaques (*Macaca nemestrina nemestrina*) in West Sumatra, *International Journal of Primatology* 11 (1990): 339–356; T. Oi, Population organization of wild pig-tailed macaques (*Macaca nemestrina nemestrina*) in West Sumatra, *Primates* 31 (1990): 15–31.

5. Other field studies include J. Caldecott, An Ecological Study of the Pig-tailed Macaque in Peninsular Malaysia (PhD thesis, Cambridge University, 1983); and J. G. Robertson, On the Evolution of Pig-tailed Macaque Societies (PhD thesis, Cambridge University, 1986). Other general studies on power in pig-tailed macaques, D. L. Castles, F. Aureli, and F. B. M. de Waal, Variation in conciliatory tendency and relationship quality across groups of pigtail macaques, *Animal Behaviour* 52 (1996): 389–403; C. Giacoma and P. Messeri, Attributes and validity of dominance hierarchy in the female pigtail macaque, *Primates* 33 (1992): 181–189; P. G. Judge, Dyadic and triadic reconciliation in pigtail macaques (*Macaca nemestrina*), *American Journal of Primatology* 23 (1991): 225–237.

6. P. Maxim, Contexts and messages in macaque social communication, *American Journal of Primatology* 2 (1982): 63–85; F. B. M. de Waal and L. M. Luttrell, The formal hierarchy of rhesus macaques: An investigation of the bared tooth display, *American Journal of Primatology* 9 (1985): 73–85; J. C. Flack and F. B. M. de Waal, Context modulates signal meaning in primate communication, *Proceedings of the National Academy of Sciences* 104 (2007): 1581–1586.

7. None of these effects were found when a second experiment was run in which less powerful, low-ranking individuals were behaviorally knocked out of the troop.

8. Results from the knockout experiment are spread across two papers: J. C. Flack, D. C. Krakauer, and F. B. M. de Waal, Robustness mechanisms in primate societies: A perturbation study, *Proceedings of the Royal Society of London* 272 (2005): 1091–1099; and J. C. Flack, M. Girvan, F. B. M. de Waal, and D. C. Krakauer, Policing stabilizes construction of social niches in primates, *Nature* 439 (2006): 426–429. See also J. C. Flack, F. B. M. de Waal, and D. C. Krakauer, Social structure, robustness, and policing cost in a cognitively sophisticated species, *American Naturalist* 165 (2005): E126–E139; J. C. Flack and D. C. Krakauer, Encoding power in communication networks, *American Naturalist* 168 (2006): E87–E102.

9. V. Pallante, R. Stanyon, and E. Palagi, Agonistic support towards victims buffers aggression in geladas (*Theropithecus gelada*), *Behaviour* 153 (2016): 1217–1243; E. Palagi, A. Leone, E. Demuru, and P. F. Ferrari, High-ranking geladas protect and comfort others after conflicts, *Scientific Reports* 8 (2018), https://doi.org/10.1038/s41598-018-33548-y.

10. For more on the Australian National Botanic Gardens, see https://parks

australia.gov.au/botanic-gardens/ and http://www.anbg.gov.au/gardens/living
/gardens-profile/index.html.

11. D. J. Green, A. Cockburn, M. L. Hall, H. Osmond, and P. O. Dunn, Increased opportunities for cuckoldry may be why dominant male fairy-wrens tolerate helpers, *Proceedings of the Royal Society of London* 262 (1995): 297–303; P. O. Dunn, A. Cockburn, and R. A. Mulder, Fairy-wren helpers often care for young to which they are unrelated, *Proceedings of the Royal Society of London* 259 (1995): 339–343; R. A. Mulder, P. O. Dunn, A. Cockburn, K. A. Lazenbycohen, and M. J. Howell, Helpers liberate female fairy-wrens from constraints on extra-pair mate choice, *Proceedings of the Royal Society of London* 255 (1994): 223–229.

12. R. A. Mulder and N. E. Langmore, Dominant males punish helpers for temporary defection in superb fairy-wrens, *Animal Behaviour* 45 (1993): 830–833; A. J. Gaston, Evolution of group territorial behavior and cooperative breeding, *American Naturalist* 112 (1978): 1091–1100.

13. Smaller juvenile helpers are often the dominant pair's offspring from prior clutches. But not all juveniles help their parents. Barbara Taborsky (Michael's significant other) wanted to know why, and she and her colleagues used whole-genome sequencing to find out. In particular, they measured gene expression in the brains of fish they tested to try to piece together why helping was seen in some young, but not others. When they reached one hundred days of age, sixty-two juveniles were each presented with the opportunity to clean the eggs in a clutch, one of the more prominent helping behaviors seen at this age. Prior work had shown that individuals are very consistent about cleaning or not, so the team characterized these fish as cleaners or non-cleaners. Then, fifteen days later, they divided both cleaner and non-cleaner groups, such that within each group, some fish had another opportunity to clean, while others did not (controls).

Taborsky and her team found differences in "resting gene expression"— differences between cleaners and non-cleaners that were not due to a recent act of cleaning—in only a single gene (*irx2*). That gene is associated with basic neural development in vertebrates, especially what Taborsky and her team call "long-term organizational effects on brain morphology and physiology." When they next compared cleaners and non-cleaners after they were presented the opportunity to clean at day 115, they found seven genes that showed increased expression levels—technically called overexpression. Overexpression of three of these genes (*c-fos*, *nr4a3*, and *mb9.15*) was found in both cleaners and non-cleaners. Though very cautious in their interpretation of the overexpression of these three genes, based on what other studies have found about their function, Taborsky and her team suggest that they prime juveniles for the act of cleaning, whether they clean or not. The four genes (*Csrnp1b*, *epsti1*, *rsad2*, and *ido2*) that were overexpressed only in cleaners are part of what has been dubbed a "social toolkit," associated with the molecular and neurobiological regulation mechanism of social behavior, including helping behavior. C. Kasper, F. O. Hebert, N. Aubin-Horth, and B. Taborsky, Divergent brain gene expression profiles between alternative behavioural helper types in a cooperative breeder, *Molecular Ecology* 27 (2018): 4136–4151.

Chapter Seven

1. R. C. Connor, M. R. Heithaus, and L. M. Barré, Superalliance of bottle-nose dolphins, *Nature* 397 (1999): 571–572; R. C. Connor, J. J. Watson-Capps, W. S. Sherwin, and M. Krützen, New levels of complexity in the male alliance networks of Indian Ocean bottlenose dolphins (*Tursiops* sp.), *Biology Letters* 7 (2011): 623–626; R. C. Connor and M. Krützen, Male dolphin alliances in Shark Bay: Changing perspectives in a 30-year study, *Animal Behaviour* 103 (2015): 223–235; R. C. Connor, Dolphin social intelligence: Complex alliance relationships in bottlenose dolphins and a consideration of selective environments for extreme brain size evolution in mammals, *Philosophical Transactions of the Royal Society of London* 362 (2007): 587–602; L. Marino, R. C. Connor, R. E. Fordyce, L. M. Herman, P. R. Hof, L. Lefebvre, D. Lusseau et al., Cetaceans have complex brains for complex cognition, *PLoS Biology* 5 (2007): 966–972.

2. D. De Luca, The Socio-Ecology of a Plural Breeding Species: The Banded Mongoose (*Mungos mungo*) in Queen Elizabeth National Park Uganda (PhD thesis, University College London, 1998).

3. F. J. Thompson, H. H. Marshall, J. L. Sanderson, E. I. K. Vitikainen, H. J. Nichols, J. S. Gilchrist, A. J. Young, S. J. Hodge, and M. A. Cant, Reproductive competition triggers mass eviction in cooperative banded mongooses, *Proceedings of the Royal Society of London* 283 (2016), https://doi.org/10.1098/rspb .2015.2607; F. J. Thompson, H. H. Marshall, E. I. K. Vitikainen, A. J. Young, and M. A. Cant, Individual and demographic consequences of mass eviction in cooperative banded mongooses, *Animal Behaviour* 134 (2017): 103–112; F. J. Thompson, M. A. Cant, H. H. Marshall, E. I. K. Vitikainen, J. L. Sanderson, H. J. Nichols, J. S. Gilchrist et al., Explaining negative kin discrimination in a cooperative mammal society, *Proceedings of the National Academy of Sciences* 114 (2017): 5207–5212; L. A. Dugatkin, Long reach of inclusive fitness, *Proceedings of the National Academy of Sciences* 114 (2017): 5067–5068.

4. F. J. Thompson, H. H. Marshall, E. I. K. Vitikainen, and M. A. Cant, Causes and consequences of intergroup conflict in cooperative banded mongooses, *Animal Behaviour* 126 (2017): 31–40; R. Johnstone, M. A. Cant, D. Cram, and F. J. Thompson, Exploitative leaders incite intergroup warfare in a social mammal, *Proceedings of the National Academy of Sciences* 117 (2020): 29759–29766; M. A. Cant, E. Otali, and F. Mwanguhya, Fighting and mating between groups in a cooperatively breeding mammal, the banded mongoose, *Ethology* 108 (2002): 541–555. For more on the general dynamics of social life in banded mongooses, see H. J. Nichols, W. Amos, M. A. Cant, M. B. V. Bell, and S. J. Hodge, Top males gain high reproductive success by guarding more successful females in a cooperatively breeding mongoose, *Animal Behaviour* 80 (2010): 649–657; M. B. V. Bell, H. J. Nichols, J. S. Gilchrist, M. A. Cant, and S. J. Hodge, The cost of dominance: Suppressing subordinate reproduction affects the reproductive success of dominant female banded mongooses, *Proceedings of the Royal Society of London* 279 (2012): 619–624; M. A. Cant, H. J. Nichols, R. A. Johnstone, and S. J. Hodge, Policing of reproduction by hidden threats in a cooperative mammal, *Proceedings of the National Academy of Sciences* 111 (2014): 326–330.

5. J. Abumrad and R. Krulwich, "Argentine Invasion," July 30, 2012, in

Radiolab, podcast, https://www.wnycstudios.org/podcasts/radiolab/articles /226523-ants.

6. M. L. Thomas, C. M. Payne-Makrisa, A. V. Suarez, N. D. Tsutsui, and D. A. Holway, Contact between supercolonies elevates aggression in Argentine ants, *Insectes Sociaux* 54 (2007): 225–233; M. L. Thomas, C. M. Payne-Makrisa, A. V. Suarez, N. D. Tsutsui, and D. A. Holway, When supercolonies collide: Territorial aggression in an invasive and unicolonial social insect, *Molecular Ecology* 15 (2006): 4303–4315; A. V. Suarez, D. A. Holway, D. S. Liang, N. D. Tsutsui, and T. J. Case, Spatiotemporal patterns of intraspecific aggression in the invasive Argentine ant, *Animal Behaviour* 64 (2002): 697–708.

7. A. V. Suarez, D. A. Holway, and N. D. Tsutsui, Genetics and behavior of a colonizing species: The invasive Argentine ant, *American Naturalist* 172 (2008): S72–S84; N. D. Tsutsui, A. V. Suarez, and R. K. Grosberg, Genetic diversity, asymmetrical aggression, and recognition in a widespread invasive species, *Proceedings of the National Academy of Sciences* 100 (2003): 1078–1083; N. D. Tsutsui, A. V. Suarez, D. A. Holway, and T. J. Case, Reduced genetic variation and the success of an invasive species, *Proceedings of the National Academy of Sciences* 97 (2000): 5948–5953.

8. J. L. Fitzpatrick and B. J. Bowman, Florida scrub-jays: Oversized territories and group defense in a fire-maintained habitat, in *Cooperative Breeding in Vertebrates: Studies of Ecology, Evolution, and Behavior*, W. Koenig and J. Dickinson, eds. (Cambridge: Cambridge University Press, 2016), 77–97; A. R. Degange, J. W. Fitzpatrick, J. N. Layne, and G. E. Woolfenden, Acorn harvesting by Florida scrub jays, *Ecology* 70 (1989): 348–356.

9. G. E. Woolfenden and J. W. Fitzpatrick, Inheritance of territory in group-breeding birds, *BioScience* 28 (1978): 104–108; G. E. Woolfenden and J. W. Fitzpatrick, Florida scrub jays: A synopsis after 18 years, in *Cooperative Breeding in Birds*, P. B. Stacey and W. D. Koenig, eds. (Cambridge: Cambridge University Press, 1990), 241–266; G. E. Woolfenden and J. W. Fitzpatrick, *The Florida Scrub Jay: Demography of a Cooperative-Breeding Bird* (Princeton: Princeton University Press, 1984).

10. A. Bhadra and R. Gadagkar, We know that the wasps "know": Cryptic successors to the queen in *Ropalidia marginata*, *Biology Letters* 4 (2008): 634–637; A. Bhadra, P. L. Iyer, A. Sumana, S. A. Deshpande, S. Ghosh, and R. Gadagkar, How do workers of the primitively eusocial wasp *Ropalidia marginata* detect the presence of their queens? *Journal of Theoretical Biology* 246 (2007): 574–582; A. Bhadra, Queens and Their Successors: The Story of Power in a Primitively Eusocial Wasp (PhD thesis, Centre for Ecological Sciences, Bangalore, India, 2008).

11. M. E. Gompper, ed., *Free-Ranging Dogs and Wildlife Conservation* (Oxford: Oxford University Press, 2013); W. Bollee, *Gone to the Dogs in Ancient India* (Munich: Verlag der Bayerischen Akademie der Wissenschaften, 2006); B. Debroy, *Sarama and Her Children: The Dog in Indian Myth* (New York: Penguin Books, 2008).

12. S. Sen Majumder, A. Bhadra, A. Ghosh, S. Mitra, D. Bhattacharjee, J. Chatterjee, A. K. Nandi, and A. Bhadra, To be or not to be social: Foraging associations of free-ranging dogs in an urban ecosystem, *Acta Ethologica* 17 (2014):

1–8; D. Bhattacharjee, S. Dasgupta, A. Biswas, J. Deheria, S. Gupta, N. N. Dev, M. Udell, and A. Bhadra, Practice makes perfect: Familiarity of task determines success in solvable tasks for free-ranging dogs (*Canis lupus familiaris*), *Animal Cognition* 20 (2017): 771–776; M. Paul, S. Sen Majumder, and A. Bhadra, Selfish mothers? An empirical test of parent-offspring conflict over extended parental care, *Behavioural Processes* 103 (2014): 17–22; S. Sen Majumder, A. Chatterjee, and A. Bhadra, A dog's day with humans: Time activity budget of free-ranging dogs in India, *Current Science* 106 (2014): 874–878; A. Bhadra, D. Bhattacharjee, M. Paul, A. Singh, P. R. Gade, P. Shrestha, and A. Bhadra, The meat of the matter: A rule of thumb for scavenging dogs? *Ethology Ecology & Evolution* 28 (2016): 427–440; M. Paul, S. Sen Majumder, and A. Bhadra, Grandmotherly care: A case study in Indian free-ranging dogs, *Journal of Ethology* 32 (2014): 75–82; S. K. Pal, Factors influencing intergroup agonistic behaviour in free-ranging domestic dogs (*Canis familiaris*), *Acta Ethologica* 18 (2015): 209–220. Robert Bonanni has studied intergroup power dynamics in street dogs in Rome: R. Bonanni, P. Valsecchi, and E. Natoli, Pattern of individual participation and cheating in conflicts between groups of free-ranging dogs, *Animal Behaviour* 79 (2010): 957–968; R. Bonanni, E. Natoli, S. Cafazzo, and P. Valsecchi, Free-ranging dogs assess the quantity of opponents in intergroup conflicts, *Animal Cognition* 14 (2011): 103–115.

13. M. C. Crofoot, D. I. Rubenstein, A. S. Maiya, and T. Y. Berger-Wolf, Aggression, grooming and group-level cooperation in white-faced capuchins (*Cebus capucinus*): Insights from social networks, *American Journal of Primatology* 73 (2011): 821–833.

14. M. C. Crofoot, The cost of defeat: Capuchin groups travel further, faster and later after losing conflicts with neighbors, *American Journal of Physical Anthropology* 152 (2013): 79–85.

15. M. C. Crofoot, I. C. Gilby, M. C. Wikelski, and R. W. Kays, Interaction location outweighs the competitive advantage of numerical superiority in *Cebus capucinus* intergroup contests, *Proceedings of the National Academy of Sciences* 105 (2008): 577–581; For more on evolution and the free-rider problem, see C. T. Bergstrom and L. A. Dugatkin, *Evolution* (New York: W. W. Norton, 2016); M. C. Crofoot and I. C. Gilby, Cheating monkeys undermine group strength in enemy territory, *Proceedings of the National Academy of Sciences* 109 (2012): 501–505.

16. A. V. Jaeggi, B. C. Trumble, and M. Brown, Group-level competition influences urinary steroid hormones among wild red-tailed monkeys, indicating energetic costs, *American Journal of Primatology* 80 (2018), https://doi.org/10.1002/ajp.22757; M. Brown, Intergroup Encounters in Grey-Cheeked Mangabeys (*Lophocebus albigena*) and Redtail Monkeys (*Cercopithecus ascanius*): Form and Function (PhD thesis, Columbia University, 2011).

Chapter Eight

1. J. J. M. Massen, A. Pasukonis, J. Schmidt, and T. Bugnyar, Ravens notice dominance reversals among conspecifics within and outside their social group, *Nature Communications* 5 (2014), https://doi.org/10.1038/ncomms4679.

2. A. Guhl, Social behavior of the domestic fowl, *Technical Bulletin, Kanas*

Agricultural Experimental Station 73 (1953): 3–48; M. D. Breed, S. K. Smith, and B. G. Gall, Systems of mate selection in a cockroach with male dominance hierarchies, *Animal Behaviour* 28 (1980): 130–134; A. Moore, The inheritance of social dominance, mating behavior and attractiveness to mates in male *Nauphoeta cinerea*, *Animal Behaviour* 39 (1990): 388–397; A. Moore and M. Breed, Mate assessment in a cockroach, *Nauphoeta cinerea*, *Animal Behaviour* 34 (1986): 1160–1165; A. Moore, W. Ciccone, and M. Breed, The influence of social experience on the behaviour of male cockroaches, *Nauphoeta cinerea*, *Journal of Insect Behavior* 1 (1988): 157–168.

3. L. A. Dugatkin, M. Alfieri, and A. J. Moore, Can dominance hierarchies be replicated? Form-reform experiments using the cockroach, *Nauphoeta cinerea*, *Ethology* 97 (1994): 94–102.

4. A. V. Georgiev, D. Christie, K. A. Rosenfield, A. V. Ruiz-Lambides, E. Maldonado, M. E. Thompson, and D. Maestripieri, Breaking the succession rule: The costs and benefits of an alpha-status take-over by an immigrant rhesus macaque on Cayo Santiago, *Behaviour* 153 (2016): 325–351; D. Maestripieri, *Macachiavellian Intelligence: How Rhesus Macaques and Humans Have Conquered the World* (Chicago: University of Chicago Press, 2008); F. B. Bercovitch, Reproductive strategies of rhesus macaques, *Primates* 38 (1997): 247–263; J. Berard, A four-year study of the association between male dominance rank, residency status, and reproductive activity in rhesus macaques (*Macaca mulatta*), *Primates* 40 (1999): 159–175; J. H. Manson, Do female rhesus macaques choose novel males? *American Journal of Primatology* 37 (1995): 285–296. For a more general review, see J. A. Teichroeb and K. M. Jack, Alpha male replacements in nonhuman primates: Variability in processes, outcomes, and terminology, *American Journal of Primatology* 79 (2017), https://doi.org/10.1002/ajp.22674.

5. J. P. Higham and D. Maestripieri, Revolutionary coalitions in male rhesus macaques, *Behaviour* 147 (2010): 1889–1908.

6. J. C. Azkarate, J. C. Dunn, C. D. Bacells, and J. V. Baro, A demographic history of a population of howler monkeys (*Alouatta palliata*) living in a fragmented landscape in Mexico, *Peerj* 5 (2017), https://doi.org/10.7717/peerj.3547.

7. *Astatotilapia burtoni* is sometimes called *Haplochromis burtoni*. For some general reviews of behavior in this species, see K. P. Maruska and R. D. Fernald, *Astatotilapia burtoni*: A model system for analyzing the neurobiology of behavior, *ACS Chemical Neuroscience* 9 (2018): 1951–1962; R. D. Fernald and K. P. Maruska, Social information changes the brain, *Proceedings of the National Academy of Sciences* 109 (2012): 17194–17199; J. L. Loveland, N. Uy, K. P. Maruska, R. E. Carpenter, and R. D. Fernald, Social status differences regulate the serotonergic system of a cichlid fish, *Astatotilapia burtoni*, *Journal of Experimental Biology* 217 (2014): 2680–2690.

8. K. P. Maruska, L. Becker, A. Neboori, and R. D. Fernald, Social descent with territory loss causes rapid behavioral, endocrine and transcriptional changes in the brain, *Journal of Experimental Biology* 216 (2013): 3656–3666; K. P. Maruska, Social transitions cause rapid behavioral and neuroendocrine changes, *Integrative and Comparative Biology* 55 (2015): 294–306; K. P. Maruska, A. Zhang, A. Neboori, and R. D. Fernald, Social opportunity causes rapid transcriptional changes in the social behaviour network of the brain in an African

cichlid fish, *Journal of Neuroendocrinology* 25 (2013): 145–157; R. E. Carpenter, K. P. Maruska, L. Becker, and R. D. Fernald, Social opportunity rapidly regulates expression of CRF and CRF receptors in the brain during social ascent of a teleost fish, *Astatotilapia burtoni*, *PLoS ONE* 9 (2014), https://doi.org/10.1371/journal.pone.0096632; J. M. Butler, S. M. Whitlow, D. A. Roberts, and K. P. Maruska, Neural and behavioural correlates of repeated social defeat, *Scientific Reports* 8 (2018), https://doi.org/10.1038/s41598-018-25160-x.

9. K. P. Maruska and R. D. Fernald, Plasticity of the reproductive axis caused by social status change in an African cichlid fish: II. Testicular gene expression and spermatogenesis, *Endocrinology* 152 (2011): 291–302; K. P. Maruska, B. Levavi-Sivan, J. Biran, and R. D. Fernald, Plasticity of the reproductive axis caused by social status change in an African cichlid fish: I. Pituitary gonadotropins, *Endocrinology* 152 (2011): 281–290; K. P. Maruska and R. D. Fernald, Behavioral and physiological plasticity: Rapid changes during social ascent in an African cichlid fish, *Hormones and Behavior* 58 (2010): 230–240; K. P. Maruska, Social transitions cause rapid behavioral and neuroendocrine changes, *Integrative and Comparative Biology* 55 (2015): 294–306.

INDEX

Abril-de-Abreu, Rodrigo, 19
Adams, Eldridge, 53
aggression, 4, 15–16, 27, 32–33, 36–
37, 62, 64, 80, 91–92, 100–103,
105–8, 122–24, 128–29, 132–
33, 142, 144–45, 147–48, 150;
arginine-vasopressin (AVP) hormone, 69–70; cortisol levels, 137;
cost benefits, 24; cross-fostering
study, 69–70; forms of, 94; game
theory, 24, 50; helpers, 113, 116;
hierarchy, 22–23; power, quest
for, xii, 12, 21, 69, 110; preempting of, 3; sequential assessment
model, 48–49; over territory,
130–31; war of attrition, 50; winner effects, 68–69. *See also individual animals*
Alexander, Richard, 79
Alfieri, Michael, 140–41
Ang, Tzo Zen, 20–23
Animal Farm (Orwell), 117
Año State Park, 5–8, 10, 155
anthropology, 4
ants, power dynamics in, xiv
Anubis, 86
Archbold Biological Station, 125–26
Archie, Elizabeth, 28
Argentina, 121–23
Argentine ants (*Linepithema humile*),

130; aggression in, 122–24; chemical scent, as national emblem,
124; chemical signatures, 124–25;
genetic variation, 124; intragroup
power struggles, 120–24; large
supercolony (LSC), 123–24; supercolonies of, 121, 123–24
Arnhem Zoo, 91–92
Atlanta (Georgia), 142
Aubin, Thierry, 56
audience effects, 94, 152; as extrinsic
effects, 65
Austad, Steve, 50–51
Australian National Botanic Gardens,
110

baboons: aggression in, 100, plate 5;
power dynamics in, xiv
banded mongoose (*Mungos mungo*),
130, 154; in groups, 119; mass
evictions, 118–19; mechanical
drones, use of, 120, plate 6; power
dynamics, 117–20; pup care, 32;
pup survival rates, 120; reproductive suppression, 32, 118
barking deer (*Muntiacus muntjac*), 41
Barrette, Cyrille, 40–43
Barro Colorado Island (BCI), 131–32
Becker, Elizabeth, 70
Bell, Matthew, 31–34